Introduction to Statistics for the Social Sciences

Audrey J. Weiss   **Laura L. Leets**
University of California, Santa Barbara

McGraw-Hill, Inc.
College Custom Series

New York St. Louis San Francisco Auckland Bogotá
Caracas Lisbon London Madrid Mexico Milan Montreal
New Delhi Paris San Juan Singapore Sydney Tokyo Toronto

McGraw-Hill's College Custom Series consists of products that are produced from camera-ready copy. Peer review, class testing, and accuracy are primarily the responsibility of the author(s).

Introduction to Statistics for the Social Sciences

Copyright © 1994 by McGraw-Hill, Inc. All rights reserved. Printed in the United States of America. Except as permitted under the United States Copyright Act of 1976, no part of this publication may be reproduced or distributed in any form or by any means, or stored in a data base retrieval system, without prior written permission of the publisher.

1 2 3 4 5 6 7 8 9 0 PCP PCP 9 0 9 8 7 6 5 4

ISBN 0-07-069207-6

Editor: Julie Kehrwald

Cover Design: Maggie Lytle

Printer/Binder: Port City Press

*I would like to dedicate this book to Jon
and my parents.*
... Audrey J. Weiss

I would like to dedicate this book to my parents.
... Laura L. Leets

Forward

We are indebted to the faculty, staff, and students in the Communication Department at the University of California, Santa Barbara for their support during the development of this book. In particular, we appreciate Edward Donnerstein's and Pam Gibbons' invaluable advice and consistent encouragement throughout every stage of this book.

We also want to thank a number of teaching assistants at the University of California who have helped us in pretesting many of the homework exercises and examples: Cathy Boggs, Kennan Bridge, Jon Busch, Tim Cole, Susan Fox, Jake Harwood, Robin Gurien, Dolly Imrich, Laura Jansma, Patrick O' Sullivan, Stacy Smith, and Angie Williams. Additionally, we are grateful to the numerous Communication undergraduate students at UCSB who have used portions of this material and provided useful feedback that has since been incorporated into the present book.

A special thanks is extended to our superb editor, Julie Kehrwald, for her invaluable assistance in the design and implementation of the book, and to our talented graphic artist, Chris Siwinski.

CONTENTS

CHAPTER 1 Introduction to Statistics 1
- importance of statistics - qualitative and quantitative methods
- descriptive and inferential statistics - addressing math-phobia

CHAPTER 2 Collecting and Measuring Data 11
- research methods - independent and dependent variables
- levels of measurement (nominal, ordinal, interval, ratio scales)

CHAPTER 3 Describing Data 25
- distributions - graphic representations (frequency polygon and histogram) - normal distribution - skewed distributions
- measures of central tendency (mode, median, mean)
- measures of dispersion (range, standard deviation, variance)

CHAPTER 4 Working with Distributions 53
- sample distribution - population distribution - sampling distribution - distribution of differences

CHAPTER 5 Hypothesis Testing and the z-Test 87
- alternative (research) and null hypothesis - probability
- hypothesis testing (one- and two-tail) - power
- type I and type II error

CHAPTER 6 The t-Test 129
- conceptual foundation for the t-test - computational example

CHAPTER 7 Single-Factor Analysis of Variance 155
- t-test and ANOVA comparison - conceptual foundation for the single-factor ANOVA - computational example

CHAPTER 8 Multiple-Factor Analysis of Variance 185
- single- and multiple-factor ANOVA comparison - conceptual foundation for the multiple-factor ANOVA - computational example

CHAPTER 9   Correlation   225
▨ conceptual foundation for simple correlation ▨ computational example ▨ partial correlation ▨ multiple correlation

CHAPTER 10   Regression   253
▨ conceptual foundation for linear regression ▨ standard error of the estimate ▨ computational example

CHAPTER 11   Chi-Square Analysis   265
▨ parametric and nonparametric tests ▨ conceptual foundation for chi-square ▨ computational example

APPENDIX A   Tables   291
▨ critical values of the t-distribution ▨ critical values of the F-distribution ▨ critical values of the r-distribution ▨ critical values for the χ^2-distribution

APPENDIX B   Answers to Homework Exercises   299

CHAPTER 1
INTRODUCTION TO STATISTICS

Why Should I Care About Statistics?

If you begin this book questioning the worth of learning statistics, you are not alone. Many students wonder why they must take a statistics course, especially if they don't anticipate either attending graduate school or conducting research. Additionally, it is common to reason that if you wanted to learn statistics you would have majored in a discipline emphasizing math in the first place. Arguably, however, there are three good reasons why gaining an understanding of statistics is a worthwhile investment of your time and energy.

Academic Career as a Student. When you enroll in courses within the various social science disciplines you will read research reports and journal articles to satisfy course requirements and write papers which will require that you can read and conceptually understand statistics. The more you know about statistics, the easier it will be for you to critique and interpret research findings, which in turn, inevitably will increase the quality of your work at school.

Critical Thinking. Higher education is in the business of training you to think critically and analytically, that is, to train you for citizenship and work: "The good life in a democratic society...seems to rest fundamentally on one's ability to think critically about the problems with which one is confronted. The essence of the creed is that each person possesses potentialities for discovering their own problems and for developing personally satisfying and socially acceptable solutions to them...".[1] Statistics and research methods provide an avenue through which you can solve the abstract and practical problems of life.

---

[1] Dressel, P. and Mayhew, L. (1954). *General Education: Explorations in Evaluation.* Washington, D.C.: American Council on Education, p. 35.

More specifically, mastering the logic of statistics can help you to become a better problem solver, think more clearly, weigh alternatives, and recognize personal biases.

Consumer of Research. Perhaps the most important reason to learn statistics is the fact that all of us are consumers of research. Even as you enter the world of work you will most likely continue to read research articles to stay up to date or on the cutting edge in the career you choose. As a citizen attempting to make intelligent decisions about information, understanding statistics and research methods will also help you evaluate evidence and make decisions about important issues. On a daily basis you are inundated with statistical results. For example, the latest election poll published in your voting precinct shows that the Republican candidate was favored with 53% of the vote and the Democratic candidate held 47% of the vote with a margin of error of plus or minus (+/-) 4%. The American Medical Association introduces a new cure for travel sickness after a study using 50 volunteers showed that those who used the new medication were significantly less likely to experience motion sickness than were those who used other types of medication. A Roper poll reports that one fourth of Americans believe the Nazi Holocaust never happened. Eight out of ten dentists surveyed recommend *SMILE* toothpaste for their patients who want a brighter, cleaner-looking smile. A national phone poll conducted by *People Magazine* asks for subscribers' opinions regarding the current state of Prince Charles and Princess Diana's relationship. A local senator urges you to vote against the death penalty because statistical findings indicate that states implementing the death penalty actually have higher murder rates than states that do not have the death penalty.

As a consumer of research how do you know which statistical conclusions to believe and which are deceptive? Statistics can provide information that is both useful and misleading. For instance, do 80% of all dentists really recommend *SMILE* toothpaste, or do 8 out of the 10 dentists the researcher chose to interview recommend *SMILE*? Do one fourth of

Americans truly believe the Holocaust never happened or was the question on the survey poorly worded? Why do states with the death penalty have higher murder rates? Does implementing the death penalty result in more murders or do states with more murders implement the death penalty as a deterrent? Or, perhaps, the death penalty has no association with murder rates and is purely a function of the population size.

Lastly, consider the following example: a researcher claims a social intervention program reduced the number of inner-city high-school drop outs. On what basis can the researcher make this assertion? How does the researcher know that the numbers are not really due to chance? That is, how does the researcher know that the social intervention program really keeps youths in school? These are the types of questions intelligent consumers are able to answer and that *you* too will become properly equipped to answer as you read this book. Our goal is that you will approach statistical results with skepticism and a concern for alternative explanations.

Asking Questions

Keep in mind that at the most fundamental level, research is about finding answers to questions. By nature we are all curious about the world around us. Granted, our levels of curiosity may vary (e.g., with topic and age), but at one time or another we all have been driven by the spirit of inquiry. Our curiosity is articulated through questions. The strategy we use to answer those questions will largely be governed by the questions we ask. There are two general approaches to finding answers to questions.

The first type of data gathering technique, the **qualitative method**, represents information that cannot be subjected to numerical measurements. While qualitative research covers a wide range of approaches, it tends to focus on one or a small number of cases and entails intensive interviews with people or in-depth analysis of historical materials. The data analysis involves

disadvantages - does not explain cause and effect relationship; does not apply to general population; does not use random sampling.

recording observations and interpreting patterns in these data. Often this work is associated with a specific event (case studies) and attempts to provide a description of how or why the event came about. Examples could include investigating the collapse of the Soviet Union or the Chinese government violations of human rights after the Tiananmen Square demonstration. Qualitative research tends to be used for exploratory or descriptive analysis. A researcher may often choose this approach when looking for an in-depth understanding of a phenomenon, when attempting to become familiar with an area, when working in natural situations, and when seeking a new perspective.

On the other hand, **quantitative research** represents *quantities* of some phenomenon and involves measuring data on a numerical scale (e.g., weight, length, money). The numerical data, such as number of voters who indicate they will vote for Clinton or Bush, or number of minutes to complete a puzzle task, can be subjected to various statistical analyses which, in turn, provide evidence for or against hypothesized relationships. Quantitative methodologies often move beyond description to explanation, looking for causes and reasons. A researcher employs quantitative research when s/he wants to determine the accuracy of a theory, provide evidence to support or refute an explanation, build and extend theories, and investigate knowledge about an underlying process. Surveys and experiments are examples of quantitative research methods and we will talk briefly about these and other quantitative data collection techniques (research methods) in the next chapter.

Whereas some people may view qualitative and quantitative methodologies as competitive approaches, many scholars prefer to view them as complementary approaches. In line with the latter opinion, we believe that neither quantitative nor qualitative research is superior rather, the more appropriate research method is dependent upon the type of question the researcher asks. Both traditions advance knowledge about the social world, they simply differ in their style and techniques for discerning this knowledge. Frequently research combines features of each approach. In this book we will focus on analyzing data collected through quantitative methods. You will discover shortly that not only does the research question determine the general approach to doing research, (qualitative or quantitative), but also with quantitative analyses, the research question will influence which specific statistical analysis is conducted.

So, What Exactly Are Statistics?

Beginning with the grand picture, **statistics** is a set of tools that researchers can use to answer questions from numerical data. In other words, statistics is a mathematical model for reasoning. In particular, there are two general types of statistics.

When one or two numbers are used to organize and summarize a set of data in a form that is easy to comprehend they are referred to as **descriptive statistics**. All research involves the goal of description at some level. For example, you probably use the mean of a group of test scores as a way to describe how everyone in your class performed on a particular exam. In Chapter 3 we will discuss in detail two types of descriptive statistics: measures of central tendency and measures of dispersion.

Inferential statistics go beyond simple description to make inferences about a population based on data from a sample. Usually it is impossible or impractical for a researcher to gather data from an entire population, so instead data is gathered from a sample of the population. For example, when you are budgeting your finances for the academic school year, you may estimate your cost of living based on the prices you paid for housing, food, utilities, and fees over one month from the previous year. Based on this one set of observations (i.e., your financial costs for one month) you generalize to the following year. You examined a set of observations from a sample to find typical patterns and inferred what would be true for the year. Along these lines, suppose a county waste management division wants to discover the recycling behavior of the residents in their community. Instead of surveying the thousands of people residing in their district, the researchers can make predictions or generalizations about their community based on survey results obtained from a sample. In later chapters we will discuss how inferential statistics form a foundation for several of the quantitative analyses covered in this book.

How Difficult Will This Be If I'm Not a Mathematical Genius?

Many statistics books concentrate on complex formulas, proofs, and computation. The purpose of this book is to provide you with a conceptual foundation for many of the most common analyses used in research studies with an understanding of how these statistical techniques work. Our goal is to provide you with an understanding of the nature and logic of statistics. We will do some essential mathematics but everything will be explained clearly to you. If you feel a little intimidated about the mathematical aspects of statistics, do not worry. We will work through each of the statistical tests conceptually first, and then we will work through the problems mathematically. You only need to be equipped with a basic knowledge of arithmetics and have a calculator that computes simple addition, subtraction, multiplication, division, and square roots. Beyond that, we will instruct you in everything you need to know. We believe that once you understand conceptually *what* it is you are trying to do and *why* you are trying to do it, the math will become relatively easy.

CHAPTER 1 PROBLEMS

1. Explain the difference between qualitative and quantitative research.

2. A researcher is interested in the impact of gender and type of romantic touch on perceptions of relational commitment. The researcher first constructs a questionnaire and has a total of 200 male and female students give their perceptions of relational commitment on the basis of a variety of intimate touches. Second, out of the 200 completed surveys the researcher selects ten students and conducts in-depth interviews with the students on the same topic.
    a. How does the researcher collect data in both parts of the study?
    b. What type of method (*quantitative* or *qualitative*) does the researcher use in the first part of this study? In the second part?
    c. What do you think are some of the advantages of each method conducted?
    d. Is one data collection method superior to the other or is it more advantageous to use both methods? Why?

3. Several research scenarios are provided below. Indicate which questions illustrate a qualitative approach to data collection and which illustrate a quantitative approach.
    a. A marketing firm conducts a series of experimental studies to determine how different types of music in television commercials influence audiences' attitudinal responses to the product being advertised.
    b. A researcher wants to conduct in-depth interviews with several adolescents in a community to determine levels of HIV risk-taking and to develop ways to reduce such behaviors.

c. A researcher uses public opinion polls in order to survey a community as a means to assist a jury in determining community standards of obscenity.

d. A researcher examines the patterns of violence in abusive families by living with two different families over several weeks and keeping detailed notes on each.

4. _____ statistics use a small subset of people to infer characteristics about a much larger group of people, and _____ statistics describe overall characteristics of data.

5. For each of the following statements, indicate whether it is an example of descriptive or inferential statistics.

   a. Based on last semester's text prices, you want to predict how much you will spend on text books for the following school year.

   b. The student government wants to know the percentage of students who cast votes for each of two candidates in an election.

   c. A baseball player wants to know his batting average over the last season.

   d. A rape prevention center computes the probability of date rape in a community based on the number of reported assaults in the past ten years.

6. For each of the following, try to determine what might be *misleading* about the data that is presented. Use your critical abilities and try to generate alternative explanations or limitations. At this point, do not worry about using technical terms, just try to see if you can come up with a question or possible problem about the findings.

a. The latest issue of *Cosmopolitan* publishes the following statistical information based on the results of a write-in survey from a previous issue: 85% of women are unhappy in their relationships. Should you believe that 85% of all women in the U.S. are unhappy in their relationships? Why or why not?

b. In a court of law, defense lawyers believe that it is to their advantage to have jurors of the same race as their client. This attitude was formed after the lawyers read an article reporting the results of an investigation conducted on university students who read short sample legal scenarios and made judgments of defendants' innocence or guilt. The evaluations differed according to race with more innocent judgments coming from student jurors who were of the same race as the alleged defendant. To what extend should lawyers rely on these findings?

c. CNN asks viewers to call in and indicate whether they favor universal health care coverage. Results revealed that 60% of the callers favored it and 40% did not. Do you think these data are representative of all citizens of the United States? Why or why not?

CHAPTER 2
COLLECTING AND MEASURING DATA

Collecting Data

Before conducting statistical analyses or reporting any results a researcher first needs to collect data. Researchers have several techniques at their disposal to help them gather data, which are referred to as **research methods**. The kinds of questions or topics being addressed by the researcher determines the particular methodology chosen to gather the data. No single research methodology can be used to answer all questions raised in any particular area. In the following section we will provide a brief overview of four main research methods, three of which are quantitative data collection techniques, since the purpose of this book is to introduce you to the logic and application of quantitative statistical analyses.

Research Methods

We will begin by describing one of the most common qualitative methodologies, field research, which is also called ethnography or participation observation. Field research involves the study of people in natural settings; essentially it is a sophisticated form of people watching. The researcher enters into the daily routine of some group (e.g., homeless people, gangs, cults, detectives) and attempts to describe behavior as it naturally occur in a real-life setting. While taking detailed notes on a regular basis, the researcher may directly talk with and observe people. The goal is to acquire an insider's perspective while simultaneously maintaining the distance or objectivity of an outsider. When a researcher wants to learn something that can only be studied through direct involvement and close, detailed observation, field research should be used to gather data. Field research generally is valuable for exploratory and descriptive studies. For example, one researcher desired to investigate the impact of conservatism and antifeminism on nine organizations of NOW (National Organization of Women). The researcher obtained access to NOW, their documents, and interviewed past and current members.[2] In another example of field research, researchers conducted interviews with the management and staff at Disneyland to examine the impact of changing the root-metaphor of the organization from that of "drama" to "family".[3]

The remaining three research methods we will describe, content analysis, surveys, and experiments, are all quantitative methodologies. **Content analysis** is a methodology for *analyzing* the *content* of written, visual, or spoken material. This is often referred to as a nonreactive research technique because the researcher cannot influence what is being studied. The researcher identifies the material to analyze (e.g., violent acts on television) and then creates a system or categories for recording it (e.g., verbal violence, physical violence). The coding process is used to turn content (that represent variables) into numbers that can be subjected to statistical analysis. Content analysis tends to be used in both descriptive and explanatory research. For example, researchers interested in the sexual content of music videos conducted a content analysis of a sample of television music videos videotaped over several weeks[4]. The videos were coded into five categories of physical intimacy: flirtation, non-intimate touch, intimate touch, hug, and kiss. Results revealed that sex in music videos tended to be more implied (non-intimate touch, flirtation) than direct (kiss, hug, intimate touch).

Surveys are the most frequently conducted of all quantitative research methods. More than likely you have filled out several surveys during your

---

[2] Hyde, C. (1994). Reflection: A research story. In C.K. Riessman (ed.) *Qualitative Studies in Social Work Research*. Ca: Sage Publications.

[3] Smith, R. and Eisenberg, E. (1987). Conflict at Disneyland: A root-metaphor analysis. *Communication Monographs, 54*, 367-380.

[4] Sherman, B. and Dominick, J. (1986). Violence and sex in music videos: TV and rock 'n' roll. *Journal of Communication, 36*, 79-93.

lifetime (in the mail, on the phone, maybe even you have been personally interviewed by someone). The survey is a process in which researchers translate research problems into items on questionnaires. The answers provided by respondents to the various questions can be statistically analyzed. The objective of survey research may be either explanatory or descriptive. Often researchers want to describe some characteristic about a population.

A **population** is the large group of people (or objects, such as TV shows, books, legal records, etc.) that you are interested in. For example, political researchers might be interested in all American voters. Child psychologists might be interested in the population of children. Sociologists might be interested in the population of urban youths. Communication scholars might be interested in all violent programs on television. Most of the time populations are too large to reasonably survey every single member or item of that population. Imagine, for example, if the Nielsen Company were required to solicit the viewing habits of every single individual in America, or if the Gallup polling service had to ask every American who he or she was going to vote for in the presidential election! However, even though obtaining data from all members of a population may be an insurmountable goal, researchers still may wish to make statements or inferences about the larger population. The best way to accomplish this without having to survey every member of a population (a near-impossible task, at best) is to take a subset or **sample** of the population of interest and survey the members selected for the sample.

Typically, samples selected for survey research are random or representative samples. A **random sample** is chosen in such a way that data collected from those members of the sample are representative of all members of the population. A random sample typically involves giving every single member of the population an equal chance of being included in the final sample. Which members of the population that eventually are chosen to be in the final sample can be selected through a variety of random methods including a coin toss, table of random numbers, or by selecting names out of a hat. With random sampling, the statistical results you obtain from the sample members can be generalized to the larger population of all members. For instance, political researchers inevitably sample only 1,000 or 2,000 people regarding their voting behavior. But the statistics from these few people are generalized to the entire population of voters, such that at any point during an election we know the percentage of the voting population (within some margin of error due to sampling) that is going to vote for each candidate.

Another popular quantitative methodology is the **experiment**. As a college student, it is also possible that you have been asked to participate in a research experiment as a part of a course you are taking. The experimental procedure was adopted by the social sciences from the natural sciences and is chiefly aimed at discovering cause-and-effect relationships. Thus, experiments are restricted to questions in which the researcher has control over the situation and can manipulate some aspects of the situation. Usually experiments involve dividing respondents into two or more groups where one group of people receive a treatment or stimulus of some sort (the experimental group) and the other group receives no treatment (the control group). Typically people are randomly assigned to groups in an experiment. Be sure to notice that random assignment is different than random sampling. **Random assignment** refers to how participants in a study (the sample) are assigned to the groups in a study. For example, if a researcher is interested in whether a new drug can cure attention deficit disorder, one way to assign people to groups is through random assignment. With a sample size of sixty volunteers (which is not a random sample) each participant can be randomly assigned to a condition (drug vs. placebo). Using a coin toss procedure, if the coin is heads the subject is assigned to receive the new drug (experimental condition) and if the coin is tails the participant is assigned to receive the placebo (control condition). Random assignment is a characteristic of true experiments because it ensures that any observed differences between groups in the study (e.g., the experimental group and the control) can be explained only by the researcher's manipulation (e.g.,

independent variable — one by which the groups are defined
dependent variable — one about which comparisons are made

type of drug -- real vs. placebo). Note that random sampling and random assignment are independent concepts. Random sampling is a method for obtaining a sample that yields representativeness. Random assignment is a process for assigning people, who may or may not be a random sample, into groups in an experiment.

In another example of an experiment, a psychologist might be interested in the extent to which positive reinforcement affects performance at a task. The psychologist might randomly assign one group of participants in the study to receive positive words of encouragement while completing a difficult puzzle task, whereas another group of participants does not receive any encouragement at all while completing the task. The researcher could then measure how long it took for each subject to complete the puzzle task, in order to determine whether those who received reinforcement were able to accomplish the task more quickly than those who did not receive reinforcement. In this study presence of positive reinforcement was the **independent variable** that the researcher manipulated. In general, a **variable** is anything that can take on multiple values. In this case, the independent variable had two values or **levels**: presence of positive reinforcement, absence of positive reinforcement. The result or outcome of the study was measured by the **dependent variable**, which in this case was number of minutes required to complete the puzzle task. This dependent measure was free to take on any value between zero and infinity. The *data* gathered in this study is through the dependent variable. Statistical tests, then, are performed on these data.

Levels of Measurement

A researcher cannot gather data and test hypotheses without measures. **Measurement** can be defined as the process of assigning numbers to objects in such a way that properties of the objects are reflected in the numbers themselves. Thus, when the numbers are manipulated (i.e., statistically), the researcher can obtain new information about the phenomena under study. However, controversy abounds when we try to determine what type of empirical facts are actually represented by this approach to measurement[5]. While there are many ways to categorize and assign measurement we will introduce you to the four scales of measurement that have the widest acceptance in the social sciences (see Table 2.1). Not only do these levels of measurement provide different amounts of information but they also serve as one important criterion in determining which statistical test may be used for analyzing data. The nominal, ordinal, interval, and ratio scales are categorized by their precision. The nominal scale provides the least amount of information and the ratio scale offers the most amount of information. Notice when you move sequentially from the nominal to the ratio scale, each scale contains the same information as the previous scale(s) while adding a piece of new information.

Scale	Characteristic of Scale
NOMINAL	Numbers represent categories; a classification scale.
ORDINAL	Numbers indicate rank order of observations. Gives the order of the numbers but not the differences between them.
INTERVAL	Known and equal intervals with an arbitrary zero point.
RATIO	Numbers represent equal units from an absolute zero point.

Table 2.1 Levels of measurement

With a **nominal scale** numbers are used simply to classify or categorize objects, people, or characteristics. No ranking is possible; numbers simply are assigned to distinguish categories from one another. Examples of nominal-level measurement are a blood bank classifying a person's blood type as O, A,

---

[5] For a good summary of various theoretical approaches to measurement, we recommend: Mitchell, J. (1986). Measurement scales and statistics: A clash of paradigms. *Psychological Bulletin, 100,* 398-407.

B, AB, and the Census Bureau gathering demographic information such as gender (male, female), religion (Protestant, Catholic, Buddhist etc.) and ethnicity (Anglo, Black, Hispanic, Asian etc.).

With an **ordinal scale**, the data are ordered or ranked. However, the intervals between these observations are undefined and unequal. For example, consider houses ordered on a street. When someone gives you directions to a house on this street, s/he may tell you that it is the third house on the left. From this, you know only the order or rank of this house on the street, not how far away it is from the first, second, or fourth houses on the street. When you go to find this house, you may find that the interval between the first and second houses is 100 feet but that the difference between the second and third houses is 1,000 feet (a somewhat longer drive).

As another example, consider finishing places in a race: Bob, Cindy, Heidi, Greg. You know Cindy finished behind Bob, but did she finish one second, one minute, 5 minutes, 20 minutes, or one hour behind Bob? And, there is nothing to indicate that this interval is the same between all runners. Whereas, Cindy may have finished 5 seconds behind Bob, Heidi may have finished 45 minutes behind her. There is no information in an ordinal-level scale regarding the interval between ranked data points (e.g., houses, runners).

When the intervals between ranks are meaningful and equal we have an **interval scale**. With this scale we now know the distance between values -- how much greater than and how much less than one value is from another. The interval scale also has an arbitrary zero point; that is, it does *not* have a meaningful zero point. Zero does not imply the absence of what we are measuring. For example, the Gregorian calendar date 0 A.D. does not represent the absence of time. Rather, it is a reference point (to the birth of Christ) in a system developed for ordering the construct of time. Other common examples of an interval scale include IQ scores and Celsius and Fahrenheit temperature scales. Hypothetically, if the temperature in Los Angeles was a pleasant 80º F during October and in San Francisco a chilly 40º

F, we would know how much warmer (i.e., 40 degrees) Los Angeles was than San Francisco but we could not say that Los Angeles was twice as warm as San Francisco. It is also important to note that the zero point on both the Celsius and Fahrenheit temperature scales is arbitrary. They do not represent the absence of temperature.

On a **ratio scale**, we obtain the most information: the measurements are ranked, intervals between the ranks are known, equal, and there is an absolute (meaningful) zero point. With a ratio scale we can determine the degree of difference between two numbers. For example, if you had a savings account of $2,000 and your roommate had a savings account of $500, we can say that you have 4 times more money saved than your roommate. Notice that a zero on the ratio scale implies the absence of what we are measuring. Thus, a zero on your banking statement would imply that you are out of money and most likely bouncing checks. Moreover, with the Kelvin temperature scale, 0 K corresponds to the absence of temperature (and life). Other common examples of ratio scales include mass or weight, volume, height, and distance.

In the beginning of this section we acknowledged that there are multiple theories to data measurement, some accepting and others rejecting the four levels of measurement. Even within this typology of scaling there is also an internal debate between researchers regarding which level of measurement is applicable to certain types of data. For example, let's say you won two promotional tickets to director Steven Spielberg's newest film release on its opening night. As you leave the theater with your date a newspaper reporter asks you to rate the movie on a Likert-type scale from 1 (very poor) to 10 (superior).

very poor   1   2   3   4   5   6   7   8   9   10   superior

How would you classify data collected on this scale? Is it ordinal or interval? We definitely have numbers in a rank order but do we have known and equal intervals? That is, are the difference between ratings of 3 and 4 the same as the difference between ratings of 8 and 9? The more conservative approach is to treat this as an ordinal-level scale, which limits the types of statistical tests you can conduct. The more common, and perhaps more practical approach, is to consider Likert-type scales as interval-level data which allow for more powerful analyses. *(can be used as ordinal level)*

One final point regarding the numbers resulting from data collection. Remember that they can often be classified according to several of the four scales, depending upon how the data is gathered and used. For example, suppose the owner of a popular nightclub wants to know the income of the patrons frequenting the facility. One evening all guests are asked to fill out a survey. The results indicate the number of people at each income level: 43 lower income, 160 middle income, and 95 upper income patrons. The owner gathered at least ordinal-level data, knowing the rank-order of the categories (lower middle, upper), but not the distance between categories (or, for that matter, whether the patrons interpreted these categories in the same way). However, if the owner gathered actual monetary incomes instead the data would be ratio-level (dollars have an absolute zero and equal intervals).

Chapter 2 Problems

1. Which research method:
   a. is aimed at discovering cause-and-effect relationships?
   b. uses questionnaires to find answers to research problems?
   c. examines people in natural settings?
   d. analyzes the content of written, visual, or spoken material?

2. All American voters can be classified as a _____, (*population* or *sample*) whereas 1,000 randomly selected American voters can be called a _____ (*population* or *sample*).

3. An educational researcher is interested in discovering whether using computers in the classroom facilitates elementary school children's' ability to read. Thirty elementary classrooms that required children to work on computers were identified and these children's scores on national reading exams were found to be higher than those of the average child in elementary school. What is the population and sample of interest?

4. In a _____ sample, every member of the population has an equal chance of being selected.

5. Which of the following are random samples (indicate all that apply):
   a. A researcher wants to know how college students at a particular university feel about a writing requirement. All students in an English course are given a questionnaire to assess their opinion about this requirement.

b. The researcher also wants to know how university professors feel about this same writing requirement. A list of all currently employed professors are obtained from the administrative building. All of the professors' names are written on a piece of paper and thrown into a hat. Twenty professors are selected from the hat to receive a questionnaire assessing their opinion of the writing requirement.

c. You want to analyze the gender of celebritys' pictures on the cover of TV Guide over the past ten years. You take a coin and flip it for every week's TV Guide and analyze only those that were present on a "tail" flip.

6. What is the difference between random sampling and random assignment?

7. Sixty communication students volunteered for public speech training and were randomly assigned to two groups. The experimental group completed a set of exercises that were designed to increase public speaking ability. The control group spent the same amount of time discussing why they had volunteered for the training. Both groups of students then gave short speeches in front of a class and were evaluated. Identify the independent and dependent variable(s).

8. A fashion designer wanted to know if overall sales of his line of clothing would be different between a major department store vs. a specialty store. The percentage of sales were gathered from both outlets every month for a year. The fashion designer then compared profits to see where the most money was grossed. Identify the independent(s) and the dependent variable(s).

22 COLLECTING AND MEASURING DATA

9. In one class 15 students are randomly assigned to receive extra help (experimental group) whereas the other 15 students are randomly assigned to take the course as in previous years with no extra help (control group). Six weeks later all of the students take the midterm exam and the researcher compares the two groups' test scores to see who performed better. Identify the independent variable(s) and the dependent variable(s).

10. Which scale of measurement:
    a. has an arbitrary zero point and equal intervals between data points?
    b. classifies people into categories?
    c. rank orders people?
    d. has an absolute zero point and equal intervals between data points?

11. Order the four levels of measurement (*ratio, nominal, ordinal,* and *interval*) from the least to the most informative. What makes each scale more precise than the one preceding it?

12. A researcher wants to measure your knowledge of the material in this course. How might the researcher measure this phenomenon? What scale of measurement is this?

13. For each of the following variables, identify the appropriate scale of measurement.
    a. nationality
    b. age
        i. in years (0 to 100+)
        ii. baby, child, adolescent, adult, senior

c. outcomes of a high school track race
    i. times (minute and seconds)
    ii. places (1st, 2nd, 3rd)
    iii. team names of competing schools
d. gender
e. Kelvin temperature
f. blood type
g. socioeconomic status
    i. in dollars
    ii. income status (lower, middle, upper)
    iii. home ownership (yes, no)
h. weight (pounds/kilograms)
i. religious affiliation
j. top 10 songs in the U.S.
    i. frequency played on the radio
    ii. list of the records
    iii. profit made off album purchases
k. centigrade temperature

CHAPTER 3
DESCRIBING DATA

Describing Data

As mentioned previously, statistics come in two general varieties: descriptive statistics, which are ways to describe data, and inferential statistics, which involve trying to make general statements for a population regarding some phenomenon based on the data you have collected from a sample (we will talk more about inferential statistics in Chapter 5). In this chapter, we will concentrate on descriptive statistics or ways to describe quantitative data once it is collected.

Suppose you are in a class with 20 students and take a midterm exam that is graded on a scale from 0 to 100. What level of measurement is this? Interval! The class period following the exam, the professor informs the class that the 20 students obtained the following scores on the exam: 87, 81, 79, 85, 85, 98, 79, 86, 91, 69, 81, 85, 86, 86, 78, 85, 91, 78, 87, 81. This is somewhat informative, but it is hard to get a good sense of what is going on with these data. Imagine if there were 100 or 1,000 scores here. By simply looking at these data it would probably be fairly difficult for you to get a good idea for how well the class performed on this exam. Did people generally do well on the exam? How spread out were these scores? Did people either do really well or really poorly? How did most of the people do? In this chapter, we will show you some ways to answer these questions by describing data in simple ways using a single number. In fact, you already know one of these descriptive statistics because you probably ask your professor this question whenever a test is returned -- what's the average (mean) score on the test?

26 DESCRIBING DATA

Distributions of Data

Because of the difficulty in eyeballing data to get a sense for what is going on, it is most convenient to put a set of numbers (data) into some type of order. Consider the 20 test scores above. These data are not in any particular order (maybe name order of the students, but certainly no type of numerical order). But, if we order these data from low to high (or high to low) we can get a better sense of what the scores are like: 69, 78, 78, 79, 79, 81, 81, 81, 85, 85, 85, 85, 86, 86, 86, 87, 87, 91, 91, 98.

[margin note: data are presented in an orderly fashion]

This already looks a bit more helpful -- there seem to be more scores in the 80s and 90s and only a few scores below 80. When data are ordered numerically as above, from the lowest number to the highest number *or* from the highest number to the lowest number, this is called a **distribution**. The same test score data would look like this if ordered from high to low: 98, 91, 91, 87, 87, 86, 86, 86, 85, 85, 85, 85, 81, 81, 81, 79, 79, 78, 78, 69. Because these are all of the data points in numerical order (albeit from high to low this time), we still have a distribution of numbers.

Certainly, the test score data is more useful when it is in the form of a distribution rather than a random ordering of data. However, we can make these data even more useful if we look at each test score and see how many times it occurred. So, for the test score data, the score 69 occurred one time (only one of the 20 students scored a 69). The score 78 occurred twice. The score 79 also occurred twice. In this way, we can create a distribution of data (because the data will remain ordered numerically) with an associated frequency of occurrence:

Score	Frequency
69	1
78	2
79	2

81	-	3
85	-	4
86	-	3
87	-	2
91	-	2
98	-	1

Perhaps not surprisingly, this type of distribution is called a **frequency distribution** because it is a distribution of numbers with a corresponding frequency. Looking at this frequency distribution, it appears that more students scored an 85 than any other score (the most frequently occurring score, which is called, the mode, will be presented later with other measures of central tendency). Relative frequency ⇒ proportion of each category

$$\frac{Frequency}{total}$$

Graphing Distributions

Often one of the best ways to get a feel for a set of data is to take a look at the numbers visually (or graphically). Now that we have created a frequency distribution of data (where each observed score occurs once along with an associated frequency of how many students got that score), we can take this information and graph it. To graph a frequency distribution, the scores or data points are plotted on the horizontal x-axis in order from lowest to highest, and the corresponding frequencies are placed on the vertical y-axis. So, an example of a graph for test score data might have the following general format:

28 DESCRIBING DATA

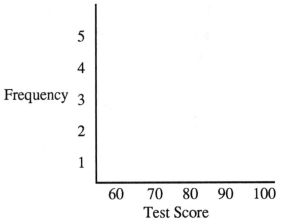

Figure 3.1 General format for plotting a frequency distribution

Now let's graph the frequency distribution of test score data for the set of 20 test scores, with one dot or point representing each score and plot one point for each score along with its corresponding frequency. The graph of these data would look like the following:

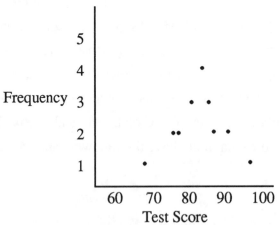

Figure 3.2 Plot of frequency distribution of test score data (exam score plotted with corresponding frequency of occurrence)

By connecting the plotted points with a line, we obtain the following graphical depiction for the 20 test scores:

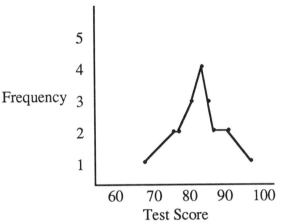

Figure 3.3 Line graph (frequency polygon) of frequency distribution of test score data

This type of graph (Figure 3.3) is known as a **line graph**. A somewhat more technical term is **frequency polygon** (polygon, meaning some type of shape, and frequency, meaning it is a plot of the frequencies of scores). The most frequently occurring score is represented by the highest point on the line graph: a score of 85 occurred with a frequency of four (that is, four people scored an 85 on the midterm exam, which was the score that occurred more frequently than any other score). The frequency polygon is particularly important, because the distributions we will be working with in the next few chapters will also be frequency polygons (used to graphically represent frequency distributions).

Another way that data sometimes are depicted graphically is by taking each data point plotted in the graph (which represents a score and its corresponding frequency of occurrence) and drawing a bar underneath with the data point at the center of each data point. Using the graph of data points from Figure 3.2, we obtain a graph like the following:

30 DESCRIBING DATA

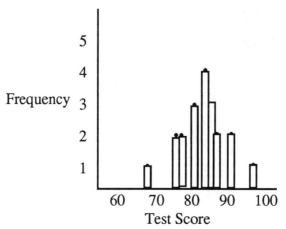

Figure 3.4 Bar graph (histogram) of frequency distribution of test score data

This type of graph is known as a **bar graph**. The name bar graph comes from the bars the fall underneath each data point. A more technical term frequently used to refer to this type of graph is a **histogram**.

Shapes of Distributions

Most graphical depictions of data, at least for the remainder of this book, will be in the form of a frequency polygon or line graph. When data are graphed in this way, the corresponding line graph can take many possible shapes. The most common shape is known as a **bell-shaped curve**, because when a line is drawn connecting the data points the resulting frequency polygon looks like a bell:

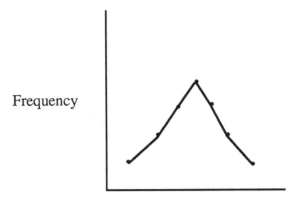

Figure 3.5 Frequency distribution in the shape of a normal distribution (normal curve or bell-shaped curve)

Many types of data in the world, such as test scores, resemble a bell-shaped curve when plotted on a line graph. Because of the commonalty of this type of data, a frequency polygon of this type is referred to as a **normal distribution** (or normal curve).

Normal distributions have a number of very important characteristics associated with them:

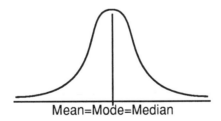
Mean=Mode=Median

1. Most of the scores cluster in the middle of the distribution
2. The normal curve is perfectly balanced or symmetrical
3. The tails will never touch the x-axis (abscissca)
4. All three measures of central tendency fall at the same point–the center
5. The normal curve has a constant relationship with the standard deviation

Table 3.1 Characteristics of a normal distribution

One of the most important characteristics of normal curves is that all of the scores cluster around the middle or the center of the distribution. Note that the most frequently occurring score, which has the highest frequency on the y-axis, is in the middle of the graph(see Figure 3.5). Another important characteristic is that the normal curve is perfectly balanced or symmetrical. If you draw a line exactly down the center (through the most frequently occurring score), the left half will be exactly the same as the right half (see Figure 3.5). So, exactly half of the scores fall to the left and half of the scores fall to the right of the center. We will see in the next chapter that this property of symmetricity is a particularly important feature of normal distributions. A third characteristic of the normal curve is that the tails of the curve never actually touch the x-axis; they are asymptotic to it (see Figure 3.5). The main reason for this asymptotic property, as we will see when we discuss these distributions in future chapters, is that normal curves actually are treated as theoretical curves. In other words, once you plot data points and create a frequency polygon, as long as it roughly represents a normal distribution, you draw a perfect normal distribution and assume that all data points would fall under this curve. Finally, there are two other important characteristics of normal distributions that are related to measures of central tendency and dispersion. We will discuss these final two characteristics when we discuss these two types of descriptive statistics shortly.

Although many types of data do resemble normal distributions, other types of data yield frequency polygons of different shapes. Two of the most common shapes for non-normal distributions are when the "tail" of one side of the distribution is spread out substantially more than the other side. When the tail on the left side of the distribution (toward the y-axis) is the longest and most spread out, the resulting distribution looks like the following:

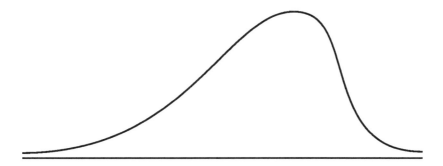

Figure 3.6 Negatively (left) skewed distribution

This type of distribution is referred to as **left** or **negatively skewed** because the longest tail is to the left or more negative end of the graph. An example of data that might produce a negatively skewed distribution is age at which people become president of a company. We would expect that relatively few children will be company presidents, slightly more young adults (early twenties) will be presidents, and we would probably expect the bulk of company presidents (the mode) to occur at around the middle or senior years of life.

When the tail on the right side of the distribution is the longest, the resulting distribution looks like the following:

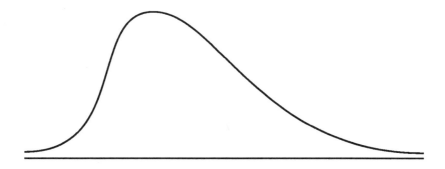

Figure 3.7 Positively (right) skewed distribution

This type of distribution is referred to as **right** or **positively skewed** because the longest tail is to the right or more positive end of the graph. An example of data that might produce a positively skewed distribution is income. Most people earn $20,000-$30,000 per year, and a few people earn $100,000 or more per year.

Measures of Central Tendency

Certainly putting data into a distribution and graphing these values makes it much easier to see what is going on with the data. It also is very helpful to describe a distribution of data with just a few summary numbers. Let's refer back to the example of 20 test scores. As a student in the class, it is probably not particularly useful to have a list of 20 scores even if they are in order (a distribution). What is more useful to you is if you know the *average* of those 20 scores, that is, a single number that describes or represents all 20 numbers. As we have suggested already, this type of summary number (the average, in this case) is an example of a descriptive statistic because it describes data. There are two general classes or types of descriptive statistics: measures of central tendency and measures of dispersion.

Measures of central tendency are descriptive statistics that try to measure the central tendency of a distribution of data, or how the data seems to cluster around the center or middle of a distribution. There are three measures of central tendency, the mean, median, and mode:

Statistic	Use
Mode (Mo)	Identifies the most frequently occurring score in a distribution. Most often used with nominal data.
Median (Mdn)	Identifies the middle score of a distribution. Most often used with ordinal data and when we do not want extreme scores to affect our measure. It can't be used with nominal data.
Mean ($\bar{X}$)	Identifies the average score in a distrubution. Only used with interval or ratio data. It can't be used with nominal or ordinal data.

Table 3.2 Measures of central tendency

More than likely you already are familiar with the mean. The **mean** of a distribution of data is simply the arithmetic average of all of the individual scores. An individual score, such as one student, Bob, who scored an 87, is represented by the symbol X. The mean of a whole group of scores is represented by $\bar{X}$ (note that some textbooks will use M to denote the mean; however, the more common convention that will be followed in this book is to use $\bar{X}$). N is used to denote the total number of scores that are included in the distribution. So, to compute the mean ($\bar{X}$), sum up (denoted by the symbol $\sum$) all of the individual scores (X) and divide by the total number of scores (N):

$$\bar{X} = \sum \frac{X}{N}$$

For example, to compute the mean for the 20 test scores:

$$\bar{X} = \frac{\begin{pmatrix} 87+81+79+85+85+98+79+86+91+69 \\ +81+85+86+86+78+85+91+78+87+81 \end{pmatrix}}{20}$$

$$\bar{X} = \frac{1678}{20} = 83.9$$

So, on average, the 20 students scored almost 84 points on this midterm exam.

Although the mean is the most common measure of central tendency, and probably the one with which you are most familiar, it is not the only way to describe the central tendency of data (nor is it necessarily the best). Suppose, for example, you are looking at income of households in a particular area. When you compute the mean you might find it to be $125,000. Wow!! People in this area are quite wealthy. Unfortunately, this conclusion is not necessarily accurate (remember the discussion in Chapter 1 about how statistics can be misleading). In fact, it may be the case that most households are only making around $30,000 a year, but that a few households are making $500,000, $1 million, and $5 million, which might be true in a place such as Hollywood. A frequency polygon of these data might look as follows:

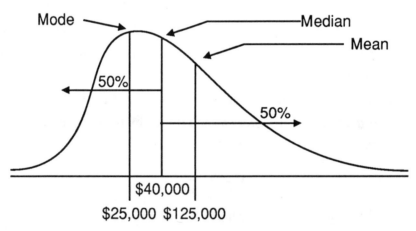

Figure 3.8 Positively (right) skewed distribution of household income data

We mentioned earlier, income is an example of a positively (right) skewed distribution. The bulk of income data appears to be clustered around $25,000, and only a few data points are at extremely high income amounts. Because of the very extreme incomes in this area (e.g., Hollywood), the mean is artificially pulled up, yielding a somewhat misleading measure of central tendency. A better way to look at the central tendency of these income data might be to find the middlemost income, that is, the income level at which 50% of the households make more than this income, and 50% of the households make less than this income. In the positively skewed income distribution (Figure 3.8) note that the median or middlemost score is $40,000; 50% of all incomes fall above $40,000 and 50% of all incomes fall below $40,000.

When you look at the middlemost score in a distribution of data, the resulting statistic is known as the **median**. The median is symbolized Mdn. Let's take the 20 test scores as an example. Which score falls in the middle of the distribution (not the frequency distribution) of these data? The distribution looked as follows:

$$\begin{array}{c} 69 \\ 78 \\ 78 \\ 79 \\ 79 \\ 81 \\ 81 \\ 81 \\ 85 \\ 85 \\ 85 \\ 85 \\ 86 \\ 86 \\ 86 \\ 87 \\ 87 \\ 91 \\ 91 \\ 98 \end{array}$$

As you can see, two numbers actually fall in the middle of this distribution. To compute the median, then, we can just average these two numbers (85 and 85). So the median of the data is $Mdn = 85$. Note that in order to compute the

38 DESCRIBING DATA

median, you must first put the data into the form of a distribution (that is, order all of the data from low to high or high to low). We cannot find the median of data that is not in the form of a distribution. For instance, consider the original 20 test scores: 87, 81, 79, 85, 85, 98, 79, 86, 91, *69, 81,* 85, 86, 86, 78, 85, 91, 78, 87, 81. The two scores in the middle are 69 and 81. However, these scores really indicate nothing about the central tendency of these data; they just happen to be the two scores that fall in the middle of a random ordering of data. Be sure to order data numerically (create a distribution) prior to computing the median.

Just as there were times when the median serves as a better measure of central tendency than the mean (e.g., with skewed distributions), sometimes neither the mean nor the median provides the best measure of the central tendency of a set of data. For instance, consider the following set of data on divorce: out of 100 married couples who eventually got divorced, 45 couples divorced after 5 years, 4 couples divorced after 10 years, 2 couples divorced after 15 years, 4 couples divorced after 20 years, and 45 couples divorced after 25 years. If we plot these data onto a graph, the resulting frequency distribution looks like this:

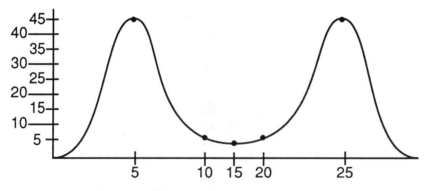

Figure 3.9 Bimodal distribution of years married prior to divorce

DESCRIBING DATA   39

Now let's compute the mean and median for these divorce data. To compute the mean we could take 5 years and put that number down 45 times or simply multiply 45 by 5 as a representation of all 45 couples who got divorced after 5 years. Similarly, we can do this for all other data points (the data point multiplied by its frequency of occurrence). So, the mean can be computed as follows:

$$\bar{X} = \frac{[(45 \times 5) + (4 \times 10) + (2 \times 15) + (4 \times 20) + (45 \times 25)]}{100} =$$

$= 15$ years of marriage prior to divorce

In this example, both the mean and the median happen to equal 15 years (note that 49 couples divorced in under 15 years and 49 couples divorced after more than 15 years and that 15 years falls in the middle of the distribution of these data). If we were to use either the mean or the median of these data we might get the impression that most couples divorce after about 15 years of marriage. However, looking at these data (see Figure 3.9), you can see that this simply is not true. In fact, most couples divorced after either a brief period of marriage (5 years) or after being married a long period of time (25 years). This statistic of the most frequently occurring score in a distribution of data is known as the **mode**. The mode is symbolized Mo. In this example of divorced couples, there actually are two modes: 5 years and 25 years.

If you look at the graph of the divorce data (see Figure 3.9) you can see that there are two peaks. These peaks represent the mode. This type of distribution is termed a **bimodal distribution** because there are two modes. Most distributions, such as the normal distribution and the positively and negatively skewed distributions, have only one mode (there is only one peak). The mode for the twenty test scores is 85; that is, 4 students scored 85, which was the most frequently occurring score (note that the peak is at 85 in Figure 3.3).

So, a particular measure of central tendency may be more appropriate depending upon how a set of data is distributed. The median is the most appropriate measure of central tendency for skewed data. The mode is the most appropriate measure of central tendency for bimodally distributed data. And, the mean is the most appropriate measure of central tendency for normally distributed data. This brings us to the fourth important characteristic of normal distributions. Remember the other three characteristics were: 1) scores cluster around the middle, 2) the distribution is symmetrical, and 3) the tails are asymptotic to the x-axis (see Table 3.1). The fourth important characteristic of a normal distribution is that all three measures of central tendency fall at exactly the same point -- the middle of the distribution. The mode is in the middle because it clearly falls at the peak. The median is in the middle because half of the scores fall above it and half fall below it, as guaranteed by the symmetrical nature of the normal curve. The mean also will fall at the middle when statistically computed. When referring to normal distributions, we almost always will refer to this middle point as the mean; however, you should not forget that the median and the mode also fall at exactly this same point.

Unlike the normal distribution, the three measures of central tendency fall at different points on skewed distributions. Recall the earlier example with household income (see Figure 3.8). We noted that a few very high incomes pulled the mean to a high value of $125,000, yielding a positively (right) skewed distribution. In contrast, the median was much lower, around $40,000. Note that the mean is the highest measure of central tendency ($125,000), the mode is the lowest measure of central tendency ($25,000), and the median is between these two ($40,000).

In contrast, what would happen with the three measures of central tendency with a negatively skewed distribution? Consider the earlier example of a distribution of data that represents the age at which someone becomes president of a company. Not surprisingly, the majority of people will probably be in their 50s or 60s. However, occasionally, a child genius might establish and become president of his or her own company. Thus, a few very low ages might skew this distribution negatively. The following is a possible graphical representation of these data:

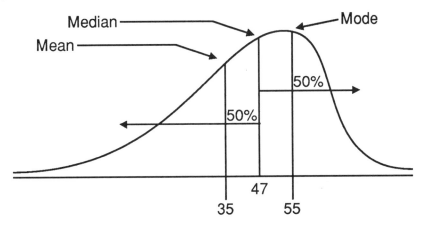

Figure 3.10 Negatively (left) skewed distribution of age of company presidents

In this case, the few low ages would pull the mean down. The mean, then, would be the smallest value on a negatively skewed distribution (say, 35 years of age). The mode would be the highest value (the most frequently occurring age of a company president, 55 years) and the median would fall in between these two (47 years).

One final note about measures of central tendency: certain measures *do not exist* with certain types of data. All of the examples in this chapter have been applied to either interval- or ratio-level data (e.g., test scores, number of years of marriage, age). However, what about data in nominal- or ordinal-level form? Let's consider ordinal-level data first. Take the Nielsen rankings of the top 20 television programs. What is the mean of these 20 shows? How about if we ask you to list your ten favorite classes in order. What is the mean of these ten classes? Clearly, with ordinal-level data the concept of a mean does not exist. Only the median and the mode are relevant concepts with data measured at an ordinal level. For example, the median of your ten favorite classes might be *Introduction to Experimental Psychology*, the class that falls in the middle of the distribution.

What about data in nominal-level form. Suppose there are 50 men and 46 women in this class. What is the mean? The median? The mode clearly is men (more men are in the class than women; men is the most frequently occurring observation or data point). Consider the class level of students at a particular university: 2,000 freshman, 2,500 sophomores, 2, 200 juniors, 1,800 seniors. What is the mean or median of these data -- a second semester sophomore? This really doesn't make sense. The mode is sophomores since the most frequently occurring class level is sophomore. With nominal-level data only the mode is a relevant concept. You cannot compute either a median or a mean with data measured at a nominal level.

Measures of Dispersion

While the three measures of central tendency are useful statistics for describing the central tendency of data, they do not provide a complete sense of the nature of the data. In particular, we might want to know something about how different the scores are from one another or how they vary. For example, suppose you know that the mean number of years a murderer spends in prison is 20 years. Does this mean that most murderers spend approximately 20 years in prison or does this mean that some murders spend only a few years in prison whereas other murderers spend 40 or 50 years in prison? The answer to this question is important for helping you understand the nature of the data.

Measures of dispersion are descriptive statistics that try to measure how far apart scores are from one another, or how spread out the data is in a distribution. There are three measures of dispersion: the range, standard deviation, and variance:

Statistic	Use
Range (R)	Identifies how far apart the lowest score and the highest scores are in a distribution.
Standard Deviation (SD)	Identifies how spread out the scores in the distribution are around the mean.
Variance (V)	Similar to standard deviation, variance is specifically the average squared deviation of scores about the mean. It is equal to the square of the SD. ($V=SD^2$) $SD=\sqrt{V}$

Table 3.3 Measures of dispersion

The **range** is simply the difference between the highest and lowest numbers in a distribution. The range is represented by the symbol R. For the set of 20 exam scores, the lowest number was 69 and the highest number was 98. So, the range, $R = 98-69 = 29$. These scores appear to be fairly spread out according to this measure. However, are most scores this far apart or is there only one person who had a 69 and everyone else had closer to a 98? This type of question might better be answered by another measure of dispersion.

The most frequently used measure of dispersion for data is the standard deviation. The **standard deviation** is a representation of how far away, on average, scores are from the mean of a set of data, and is represented symbolically as SD. For example, consider the following two sets of data for six people's weights:

Set 1		Set 2	
Joe	145	Mia	90
Mary	145	Guy	110
Bob	150	Maria	130

44 DESCRIBING DATA

> Ann   150   Ali   170
> Sue   155   Bill   190
> Tim   155   Art   210
> $\overline{X}_1 = 150$ $\overline{X}_2 = 150$

The mean for both sets of data is 150 pounds. However the weights in Set 1 appear to be much more similar to each other; that is, the weights are more homogeneous. In contrast, the weights in Set 2 are pretty dissimilar from one another (the range is a full 120 pounds; 210-90); that is, the weights are more heterogeneous. In order to get a sense of the homogeneity and heterogeneity of these two sets of data, we can look at the standard deviation. Conceptually, the standard deviation looks at how far scores are away from the mean, on average.

In Set 1, the scores appear, on average, to be close to the mean (the scores are homogenous), so the standard deviation of this distribution of data should be relatively small. Let's see how this works:

> Joe's weight of 145 is 5 pounds away from the mean of 150.
> Mary's weight of 145 is 5 pounds away from the mean of 150.
> Bob's weight of 150 is 0 pounds away from the mean of 150.
> Ann's weight of 150 is 0 pounds away from the mean of 150.
> Sue's weight of 155 is 5 pounds away from the mean of 150.
> Tim's weight of 155 is 5 pounds away from the mean of 150.

On average, the six people in Set 1 appear to weigh, on average, less than 5 pounds away from the mean, so the standard deviation will be relatively small.

What about the second group? In Set 2, the scores appear, on average, to be much further away from the mean of 150 (they are more heterogeneous), so the standard deviation of this distribution of data should be relatively large. Let's see how this works:

> Mia's weight of 90 is 60 pounds away from the mean of 150.
> Guy's weight of 110 is 40 pounds away from the mean of 150.
> Maria's weight of 130 is 20 pounds away from the mean of 150.

Ali's weight of 170 is 20 pounds away from the mean of 150.

Bill's weight of 190 is 40 pounds away from the mean of 150.

Art's weight of 210 is 60 pounds away from the mean of 150.

In contrast to Set 1, the six people in Set 2 weigh between around 40 pounds away from the mean, on average. Thus, the standard deviation for this second set of weights will be quite large. On average, the weights in Set 1 are much more homogeneous as indicated by a smaller standard deviation than are the weights in Set 2 which are more heterogeneous as indicated by a substantially larger standard deviation.

If you were to graph homogenous and heterogeneous data, the frequency polygons would look something like the following:

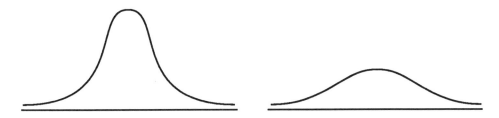

Figure 3.11 Line graphs of homogenous distribution (left) with a small standard deviation and heterogeneous distribution (right) with a large standard deviation

The graph on the left represents a very homogeneous set of data in which the standard deviation is very small. The scores are very similar to each other and no score deviates too far from the mean. In contrast, the graph on the right represents a very heterogeneous set of data in which the standard deviation is very large. The scores are very dissimilar from each other and a number of the scores are separated quite a bit from the mean.

This brings up the fifth and final characteristic of normal distributions (see Table 3.1). When a set of data is distributed normally, the distribution has a constant, fixed relationship with the standard deviation. Specifically, there are six standard deviations under the normal curve; that is, the range is approximately six standard deviations, $R \approx 6SD$ ($\approx$ means "approximately equals"); we say approximately since there are a few left over scores beyond 6SD under the tails where the curve never actually touch the x-axis. We discuss this idea of six standard deviations in the next chapter as it is an important basis for hypothesis testing and other fun things we will do in future chapters.

The third measure of dispersion or variability of data is variance. The **variance** is equivalent to the square of the standard deviation. The variance is denoted symbolically as V, so $V = SD^2$. Conceptually, variance is very similar to standard deviation as a representation of how far away, on average, scores are from the mean. Some statistical tests rely upon the variance as a more suitable measure of dispersion than the standard deviation, but conceptually the two measures are very similar.

(S^2) Sample Variance

$$\frac{\Sigma(x-\bar{x})^2}{n}$$

Sample SD (S)

$$\sqrt{\frac{\Sigma(x-\bar{x})^2}{n}}$$

nature of measures of dispersion:
① the more disperse data, the greater will be SD.
② SD is always positive
③ The more homogenous the data, the smaller will be the variance.
④ Variance is the square of the SD
⑤ If all observations have the same value, the SD will be zero.

Population Variance (σ^2)
Population SD (σ)

CHAPTER 3 PROBLEMS

1. What is a distribution?

2. When graphing a frequency distribution of scores, _____ goes on the x-axis and _____ goes on the y-axis.

3. What are more common names for a frequency polygon and a histogram? What is the difference between a frequency polygon and a histogram?

4. How does a skewed distribution differ from a normal distribution?

5. Match the verbal description with the graphical depiction for each of the following distributions:
   a. positively skewed distribution
   b. negatively skewed distribution
   c. bimodal distribution
   d. normal distribution

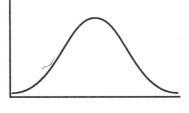
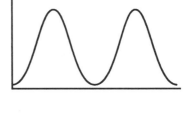

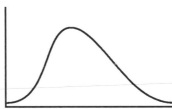

48 DESCRIBING DATA

6. Which of the following types of data would probably yield a negatively skewed distribution when the data was graphed as a frequency polygon that had a range of scores plotted along the x-axis from 0 to 100?

    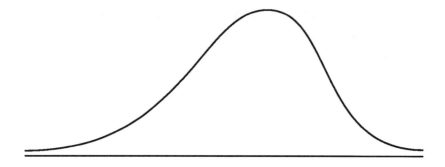

    a. number of children per American family?
    b. age at graduation from high school?
    c. scores on a very easy test?
    d. number of weekly winners of lottery tickets of more than $1 million?

7. If the mean salary for ten assistant professors in a university's department is $35,000, the mean for 15 associate professors is $45,000 and the mean for 20 full professors is $60,000, what is the mean salary for all 45 faculty members?

8. The following questions refer to the relationship between measures of central tendency and types of distributions:
    a. Which measure of central tendency is most appropriate with skewed distributions?
    b. Which measure of central tendency will have the largest value with a positively skewed distribution?

c. Which measure of central tendency will have the smallest value with a positively skewed distribution?
d. Which measure of central tendency will have the largest value with a negatively skewed distribution?
e. Which measure of central tendency will have the smallest value with a negatively skewed distribution?
f. In a positively skewed distribution, will the value of the mode be *higher* or *lower* than the value of the median?
g. In a negatively skewed distribution, will the value of the median be *higher* or *lower* than the value of the mean?
h. Which measure of central tendency is most appropriate and most frequently used with normal distributions?
i. Which measure of central tendency will have the largest value with a normal distribution?
j. Which measure of central tendency is most appropriate with a bimodal distribution?

9. The following questions refer to the relationship between measures of central tendency and levels of measurement.
   a. Which measures of central tendency can be computed when data are in ordinal-level form?
   b. Which measures of central tendency can be computed when data are in interval- or ratio-level form?
   c. Which measures of central tendency can be computed when data are in nominal-level form?

10. The following questions refer to the two distributions below:

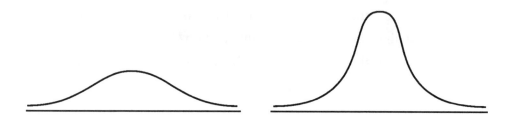

    a.   Which distribution has the largest standard deviation?
    b.   Which distribution has the smallest standard deviation?
    c.   Which distribution's scores are the most homogeneous?
    d.   Which distribution's scores are the most heterogeneous?

11. The following questions refer to two different sets of IQ scores:
    a.   If the variance of the first set of IQ scores is found to be 225, what is the value of the standard deviation?
    b.   If the standard deviation of a second set of IQ scores is 11, what is the value of the variance?
    c.   Which set of IQ score above (*a* or *b*) is the most homogeneous?

12. The following questions refer to measures of central tendency and measures of dispersion:
    a.   Which measure is the square of the standard deviation?
    b.   Which measure is the middlemost score of a distribution?
    c.   Which measure is the difference between the highest and lowest scores in a distribution?
    d.   Which measure is approximately the average distance of any given score in a distribution from the mean of the distribution?

e. Which measure provides the arithmetic average of a distribution of numbers?

f. Which measure is the most frequently occurring score in a distribution?

13. Which of the following is a distribution of numbers (choose all that apply)?

a. 78	b. 63	c. 70	d. 99	e. 73	f. 67	g. 2
69	71	94	82	81	67	6
92	79	94	73	81	82	3
83	82	82	65	81	67	3
55	88	84	65	86	77	4
76		82		87		1

14. For each of the seven sets of data listed above, compute the mean, median, mode, and range.

CHAPTER 4
WORKING WITH DISTRIBUTIONS

Types of Normal Distributions

In the last chapter, we introduced the normal distribution. The normal distribution is the most important distribution because a substantial amount of data in the real world resembles a normal distribution when graphed as a frequency polygon (line graph). In this chapter we will introduce four distributions that are all normally distributed: sample distribution, population distribution, sampl*ing* distribution, and distribution of differences (we will introduce this last distribution, the distribution of differences, in this chapter and then deal with it in detail in the next chapter). All of the characteristics of these normal distributions are very similar. For example, they all have means and standard deviations, frequency always goes on the y-axis, and the range of each distribution is equal to six standard deviations. What differs is what set of data we are actually plotting on the x-axis (e.g., individual scores, means, differences between means).

Sample Distribution

Let's start with the first two normal distributions: for a sample and for a population. If you recall from Chapter 2, a population is the entire group of people or observations that you are interested in (e.g., all students, voters, criminals, men, newspapers, etc.). A sample is a subset of the population. In most research studies the population of interest is far too large to have everyone participate, so a subset or sample of this population is selected.

Imagine, for example, that a university is offering 10 sections of an introductory statistics class and that each section enrolls 150 students. If we are interested in studying something about the population of students taking this statistics course we could either have all 1,500 students participate in the study, or select a subset of these students (e.g., 15, 20, 50 students) and have only this sample of students participate. For example, suppose we are interested in how the students in this statistics class perform on their midterm exam. If we randomly sample, say, 15 of these 1,500 students, see what each of these 15 students scored on his or her midterm (Joseph had an 86, Bob had a 79, Mary had an 82, etc.), and plot these 15 data points to form a frequency polygon, we might obtain a normal distribution that looks like the following:

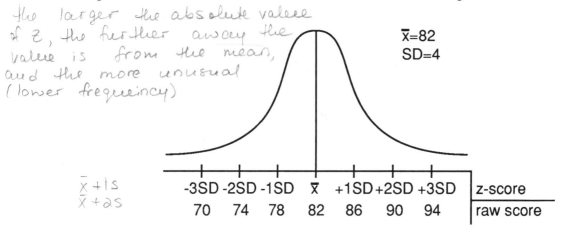

Figure 4.1 Sample distribution of test score data with mean ($\overline{X} = 82$) and standard deviation ($SD = 4$)

This resulting distribution is known as the **sample distribution** because it is a distribution of the data from a sample. The mean of these data is $\overline{X} = 82$ and the standard deviation of these data is $SD = 4$. The raw score standard deviation of a normal distribution is also referred to as the standard deviation unit of the distribution; that is, one standard deviation unit is equal to the standard deviation of the distribution. For this particular sample distribution, the standard deviation unit is 4. Since the standard deviation has a constant relationship with a normal distribution we know that approximately 6 standard deviations fit under this normal curve. That is, the range of these data is equal to 6 standard deviation units ($R \approx 6SD$). Thus, the range must equal 6 x 4 = 24. So we know that the difference between the high and low scores in this distribution is 24.

Because of the symmetrical or balanced nature of a normal distribution, we know that 3 of the standard deviation units must fall above the middle point, the mean (also, the median and mode), and that 3 of the standard deviation units must fall below the middle point. Therefore, the high point of the distribution must be 3 standard deviations above the mean or 3 x 4 = 12 points above the mean. Since the mean is 82, the highest score on this distribution must equal 94 (see Figure 4.1). Similarly, the low point of the distribution must be 3 standard deviations below the mean or 3 x 4 = 12 points below the mean. With the mean of 82, the lowest score on this distribution must equal 70. Basically, almost all of the distribution (all of the raw scores) falls between 70 and 94. We say *almost* because a very small number of scores (less than one percent, in fact) fall beyond either +3.00 SD units to the right of the distribution or -3.00 SD units to the left of the distribution. This is an artifact of the asymptotic property of the normal curve that we discussed in the previous chapter. In effect, however, you can assume that 99% of all of the scores in a normal distribution fall between a z of -3.00 and a z of +3.00 (in this example, between raw scores of 70 and 94). This is true for any normal distribution, including not only the sample distribution but also the other three types of normal distributions we will discuss in this chapter.

Since we know that the high point on the distribution is 94 and that this represents being 3 standard deviation units above the mean, we can say that a **raw score** of 94 corresponds to being 3 standard deviation units above the mean. This value of 3, or 3.00 is termed the **z-score**, which is simply a translation of a particular raw score into standard deviation units (remember in this example that one standard deviation unit equals 4). What about the z-score corresponding with a raw score of 70? We know that 70 falls 3 standard deviation units below the mean, so its corresponding z-score must be -3.00. This means that if someone scored 3 standard deviation units below the mean, they must have scored a 70.

In a similar way then, we can get a z-score representation of other raw scores in a distribution. In the sample distribution, a person who scored an 86 (raw score) had a z-score of +1.00 because s/he scored 1 standard deviation unit above the mean (see Figure 4.1). A person who scored a 74 (raw score) had a corresponding z-score of -2.00 because s/he scored 2 standard deviation units below the mean. How about someone who scored an 88 -- what would his or her corresponding z-score be? Well, since a score of 88 falls halfway between 86 and 90 (with z-scores of +1.00 and +2.00 respectively) a z-score corresponding with a raw score of 88 must be $z = +1.50$ since the raw score of 88 is $1\frac{1}{2}$ standard deviation units above the mean of 82.

This translation from raw score to z-score can work the other way too. Suppose someone scored two and a half standard deviation units below the mean (i.e. they had a z-score of -2.50). What was their raw score on this test? Since a z-score of -2.00 corresponds with a raw score of 74 and a z-score of -3.00 corresponds with a raw score of 70, then a z-score of -2.50 must correspond with a raw score of 72. How about a z-score of +0.50 -- what is the corresponding raw score? The answer is 84. How about a z-score of 0? Since a raw score of zero means the person scored no standard deviation units away from the mean, s/he must have scored exactly the mean, which is 82.

Many z-scores and raw scores can be easily computed in this way. In fact, for this particular problem you can compute all of the z-scores and raw scores in this fashion. A raw score of 83 is one fourth of a standard deviation above the mean, so the z-score is +0.25. The way in which we have been computing z-scores and raw scores can be translated into a simple formula which you may want to use when the fractions become a little too complex to compute in the way we have done above. Basically, the z-score is the difference between whatever type of score is represented on the x-axis (for the sample distribution it is a sample raw score, X) and the mean of the distribution you are working with (in this case the sample distribution has a mean of $\overline{X}$) divided by the standard deviation of the distribution you are working with (in this case the sample distribution has a standard deviation of SD):

$$z = \frac{\text{whatever is on } x\text{-axis} - \text{mean of distribution}}{\text{standard deviation of distribution}}$$

All z-scores for all distributions can be computed using this same general formula that is specialized for whatever distribution you are dealing with. That is, the z-score is how far a raw score deviates from the mean of the distribution as expressed in standard deviation units. If you are one standard deviation unit above the mean, the z-score is +1.00. If you are two standard deviation units above the mean the z-score is +2.00 and so forth. For the sample distribution the z-score formula looks like the following:

$$z = \frac{X - \overline{X}}{SD}$$

So, for example, suppose we know that Joe earned a 92 on the exam. What is his corresponding z-score? Well, because this is an easy one we can tell that Joe scored halfway between 2 standard deviation units (raw score = 90) and 3 standard deviation units (raw score = 94) above the mean, so his z-score must be +2.50. Let's see if this works using the z-score formula:

$$z = \frac{92 - 82}{4} = \frac{10}{4} = 2.50$$

The z-score formula works!

So what is the point in having z-scores to begin with? Well, z-scores are useful because no matter what type of raw scores you are working with (age, test scores, income, IQ, etc.) a z of +1.00 always refers to a score that is one standard deviation unit above the mean, regardless of the mean and standard deviation for the set of raw scores. That is, the standard deviation for a normal distribution always has a constant relationship with that distribution. This is important because it allows us to compare all sorts of distributions to each other, even if they don't have the same mean or standard deviation in their raw scores. Let's look at the example from above using 15 students' test scores so this idea will become more clear.

We noted that the mean of this distribution was 82 and the standard deviation was 4 (see Figure 4.1). Because the range always equals approximately 6 standard deviation units, we determined that the range of those test scores was 70 to 94. Suppose we obtain another group of 15 students from a statistics class the following semester and plot their scores on this same midterm exam:

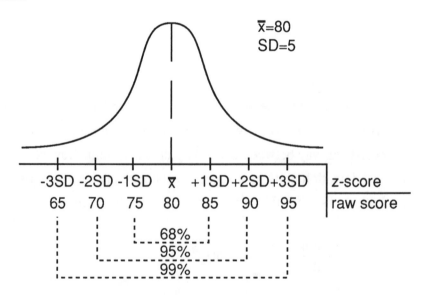

Figure 4.2 Sample distribution of test score data with mean ($\overline{X} = 80$) and standard deviation ($SD = 5$)

This time the mean is 80 and the standard deviation is 5. We know that 99% of all scores will fall between a z of -3.00 and a z of +3.00 (in this example this translates into raw scores of 65 and 95 -- the range; be sure you can work out these raw scores).

Now we can compare these two sets of students' test scores. Suppose the professor is interested in giving students from both classes who scored between 2 and 3 standard deviation units above the mean an A on the exam, between 1 and 2 standard deviation units above the mean a B, and so forth:

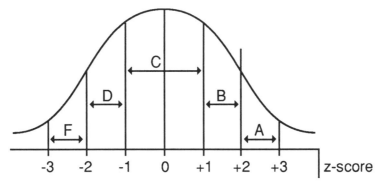

Figure 4.3 Assigning exam grades based on z-scores
(standard deviation units)

Even though the raw scores correponding to each z-score are different for the two sets of data, we can still assign grades in a similar way for the two classes by translating raw scores into z-scores. In this way, the professor can insure that she is giving the same percentage of students in each of the two classes an A, B, C, D, and F. We will see below how this use of percentages and z-scores works in more detail.

Just as there is always a fixed percentage of scores between +/-3.00 standard deviation units in a normal distribution (99%), there also is a fixed percentage of scores that fall between +/-2.00 standard deviation units around the mean. In this case, this fixed percentage is roughly 95% of all scores. That is, approximately 95% of all scores in any normal distribution will fall between a z of -2.00 and a z of +2.00. So in the earlier test score example in Figure 4.1 (with a mean of 82 and a standard deviation of 4) 95% of the students in the sample scored between a 74 and a 90. Similarly, with the second set of test score data, 95% of students scored between a 70 and 90 (see Figure 4.2). Not surprisingly, there also is a constant percentage of scores (approximately 68%) that falls between +/-1.00 standard deviations units around the mean. That is, 68% of all scores fall between a z of -1.00 and a z of +1.00. This seems to make sense. A majority of the people (over two thirds, 68%) score around the mean, where the peak is on the normal curve. You can refer to these fixed percentages as the **68-95-99 rule**: approximately 68% of all scores fall within +/-1.00 standard deviation units (z-scores) around the mean, 95% of scores fall within +/-2.00 *SD* units, and 99% of all scores fall within +/-3.00 *SD* units.

With one final piece of information, which you already know, you can determine what percentage of people fell within particular ranges of scores. Suppose we are interested in what percentage of students scored above the mean of 82 in the first set of exam score data (see Figure 4.1). Remember that the normal distribution has the property of being symmetrical (see Chapter 3); that is, the right half of the normal curve is a mirror image of the left half of the curve, so half of the scores must fall above the mean and half must fall below it. This makes sense since in a normal distribution the mean also is equivalent to the median, and the median is defined as the score at which half of all scores fall above it and half of all scores fall below it. So what percentage of students scored above the mean? The answer is 50%.

Let's combine the 50-50 symmetrical property of the normal distribution with the 68-95-99 rule and see what we can do now. Suppose the professor assigns grades based on 90 and over equals an A, 80 - 90 equals a B, and so forth, what percentage of students scored an "A" in the sample of statistics students; that is, what percentage of students scored above a 90? Well, we know that 50% scored above an 82, so fewer than 50% could have scored above 90. We also know that 95% of students scored between 74 and 90 (corresponding to a z of -2.00 and a z of +2.00). So, we can also say that 95/2 = 47.5% of students scored between 74 and the mean of 82 and 95/2 = 47.5% of students scored between the mean of 82 and 90. So, if 50% of scores are above 82, and 47.5% of scores are between 82 and 90, this must leave 50 - 47.5 = 2.5% of scores falling above a 90 (note: follow this example working from the bottom of the graph upwards)

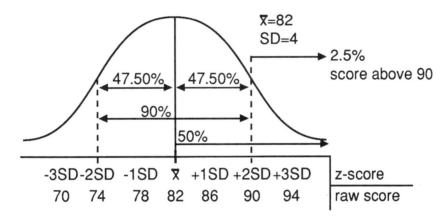

Figure 4.4 How to derive the percentage of students (2.5%) who scored above a 90

Let's try another example. What percentage of students scored below a 78? Again, we know that this percentage must be less than 50% because 50% of students scored below an 82. With the 68-95-99 rule we know that 68% of students scored between a 78 and 86 (because these are the raw scores corresponding to +/-1.00 standard deviations around the mean). So, 68/2 = 34% of students scored between 78 and the mean of 82, and 68/2 = 34% of students scored between the mean of 82 and 86. With 50% of students scoring less than 82, and 34% of students scoring between 78 and 82, then 50 - 34 = 16% of students must have scored below a 78 (note: follow this example working from the bottom of the graph upwards)

62 WORKING WITH DISTRIBUTIONS

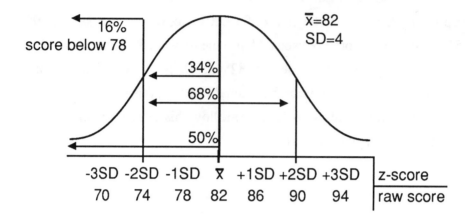

Figure 4.5 How to derive the percentage of students (16%) who scored below a 78

So far we have been talking about only one of the four types of normal distributions we introduced at the beginning of this chapter: the sample distribution. But believe it or not you now know almost everything you need to know about all four distributions! The only piece of information you are missing is what the distribution is about (that is, what type of "raw score" goes on the x-axis). You learned in this section that the raw score for a sample is plotted on the x-axis for a sample distribution.

Population Distribution

Suppose we want to look at all students' scores, not just those from one sample. In the example of the population of 1,500 statistics students, we could take all 1,500 students' scores in this statistics class and create a distribution. This resulting distribution would be called the **population distribution** because every single score in the entire population is represented. Let's assume that these test scores are distributed normally as they were with the sample of

15 students. The resulting population distribution would look like the following:

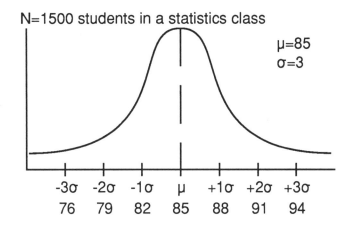

Figure 4.6 Population distribution of test scores with mean ($\mu = 85$) and standard deviation ($\sigma = 3$)

Just as with the sample distribution, the population distribution has a mean and standard deviation. The mean of this population distribution is $\mu = 85$ and the standard deviation of this population distribution is $\sigma = 3$. In this case, we refer to the mean with the Greek symbol μ (mu) and to the standard deviation with the symbol σ (sigma). Because we are referring to all scores in the population, rather than just those in the sample, we refer to parameters. A **parameter** is any type of statistic such as a mean or standard deviation that applies to all observations of a population. In contrast, the symbols $\overline{X}$ and SD are used to refer to the mean and standard deviation of a distribution of raw scores for a sample. Each of these values is termed a **statistic** because it refers to a sample of raw scores. Be sure not to confuse the use of the term *statistic* with the term *statistics*. A *statistic* is used to refer to measures like means and standard deviations that are obtained from samples rather than an entire population, whereas *statistics* is the branch of applied mathematics that you are learning about in this book.

64 WORKING WITH DISTRIBUTIONS

When discussing principles of the sample distribution we implied that you were simultaneously learning everything you need to know about the population distribution. The only piece of information you need with this second type of distribution is to know what is represented on the x-axis. As it happens, it is exactly the same thing as what was represented on the x-axis for a sample distribution: an individual raw score, X (remember that frequency is always represented on the y-axis for all distributions). You already know that distributions that are normally distributed have approximately six standard deviations under the normal curve. So, with a mean of 85 and a standard deviation of 3 (the standard deviation unit) you know that the range of this population distribution is 76 to 94 (see Figure 4.5). You also know the corresponding z-score (a score represented in standard deviation units) for each raw score on this distribution. For example, in this population distribution a raw score of 91 falls 2 standard deviation units above the mean, so this corresponds with a z-score of $z = +2.00$. What about a z-score corresponding with a raw score of 83? Well, since 83 is 2/3 of the way to 82 (which has a z-score of -1.00), 83 has a z-score of $z = -0.66$. Perhaps this one might be easier to compute using the z-score formula. You already know the general form of the z-score formula:

$$z = \frac{(\text{whatever is on } x - \text{axis}) - (\text{mean of distribution})}{\text{standard deviation of distribution}}$$

Let's apply this general formula to the population distribution. We already noted what is designated on the x-axis for the population distribution -- an individual raw score, X. What are the mean and standard deviation of the distribution we're talking about, the population distribution? As we discussed above, the mean is μ and the standard deviation is σ. So, the z-score formula for the population distribution looks like the following:

$$z = \frac{X - \mu}{\sigma}$$

In practice it works exactly the way the z-score formula did for the sample distribution. For a raw score of 83, let's use this formula to compute the z-score (we already know the answer is $z = -0.66$, let's just prove this to ourselves):

$$z = \frac{83 - 85}{3} = \frac{-2}{3} = -0.66$$

See, you already did know that.

What else do you know? Well, in the population of 1,500 students, what percentage of students scored between an 82 and 91? Well, we know that 68% of students scored between 82 and 88 (within +/- 1 SD units). Thus, the answer must be larger than 68%. We also know that 95% of all students scored between 79 and 91 (within +/- 2 SD unit). Remember the 68-95-99 rule that applies to all normal distributions. Consequently, 34% scored between 82 and 85 and 34% scored between 85 and 88 (see Figure 4.6). Since 47.5% of students scored between an 85 and 91, and 34% scored between an 85 and 88, then 47.5-34=13.5% of the students must have scored between 88 and 91. So 34% (82-85) + 34% (85-88) + 13.5% (88-91) = 81.5% (82-91) of students scored between 82 and 91 (note: follow this example working from the bottom of the graph upwards).

66 WORKING WITH DISTRIBUTIONS

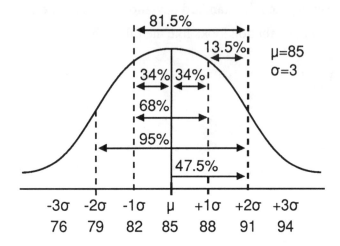

Figure 4.7 How to derive the percentage of students (81.5%) who scored between an 82 and 91

Sampling Distribution

A third important type of normal distribution differs slightly from the sample and population distributions in terms of what "raw score" goes on the x-axis. If you remember the two earlier distributions (sample and population) have a raw score X plotted on the x-axis (either an individual score for a sample or for a population, hence, the source of the names of these distributions). Rather than an individual score this third distribution has *sample means*, $\overline{X}$, plotted on the x-axis. That is, it is a distribution of sample means that is normally distributed. Let's see how this would work.

Here is the random sample of 15 students from the statistics class of 1,500 students that was used to create the sample distribution earlier in this chapter:

Student	*Test Score*
1 Joseph	86
2 Bob	79

and so forth....
15 Mary   82

Now let's compute a mean for these 15 students (we already know from earlier that for this particular sample of 15 students the mean is 82):

$$\overline{X} = \frac{86+79+.....+82}{15} = 82$$

Now let's return these 15 students to the population of 1,500 and randomly sample 15 students again. Since it is possible to re-sample Joseph, Bob etc., and Mary again at some point we call this **sampling with replacement**. That is, Joseph, Bob etc. and Mary are put back into the group (the population) from which the sample was drawn; they are replaced. This allows us to continue to take random samples of 15 people forever; that is, we are always sampling from a total of 1,500 students in the statistics class. Another random sample might look like this:

Student	*Test Score*
1 Marcia	81
2 Peter	75
and so forth....	
15 Bobby	90

Computing a mean for this sample of 15 students, we obtain the following:

$$\overline{X} = \frac{81+75+.....+90}{15} = 76$$

Now we have two sample means, 82 and 76. If we continue to do this forever we will have a whole lot of sample means. Let's say we get exhausted doing this 1,000 times and decide to graph these 1,000 sample means we've computed and see what we have:

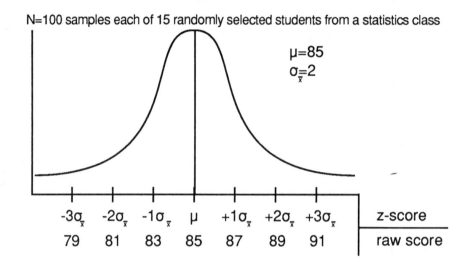

Figure 4.8 Sampling distribution of test score data with mean ($\mu = 85$) and standard deviation ($\sigma_{\bar{x}} = 2$).

This is called a distribution of sample means, or the **sampling distribution**. Notice that what we are plotting on the x-axis is not an individual score but a sample mean, $\overline{X}$. As with the other two normal distributions the sampling distribution also has a mean, μ, and a standard deviation, $\sigma_{\bar{x}}$.

Notice that the mean of the sampling distribution is symbolized μ. But wait! Isn't μ the mean for the population distribution? Yes! On the population distribution the mean is 85 *and* on the sampling distribution the mean is 85 (see Figures 4.6 and 4.8 respectively). How can exactly the same mean apply to two different distributions? Well, the population distribution includes a total of 1,500 students who constituted the population. So too does the sampling distribution, except this time all 1,500 students are represented in groups or samples of size 15. Because we sampled students 15 at a time exhaustively (or until we got exhausted) all 1,500 students eventually got represented, except each person's individual raw score got averaged in with 14 other scores to form a mean. However, because all 1,500 students are still represented on the sampling distribution, the mean of the sampling distribution has to be identical to the mean of the population distribution, $\mu = 85$.

Although the mean for both the population and the sampling distributions will be the same, their standard deviations will be different. If you notice the standard deviation for the population distribution was $\sigma = 3$ whereas the standard deviation for the sampling distribution is smaller, $\sigma_{\bar{x}} = 2$. The major reason for this is that the range of data points (on the x-axis) is smaller with the sampling distribution ($R = 12$, the difference between the endpoints of 79 and 91) than with the population distribution ($R = 18$, the difference between 76 and 94), again, see Figures 4.6 and 4.8 respectively. So why is the range smaller with the sampling distribution than with the population distribution? Consider an individual who scored a 76, which is approximately the lowest possible score on the population distribution. On the population distribution a 76 counts as a 76 and is plotted as a 76. However, on the sampling distribution this score of 76 gets averaged in with other scores when creating a sample mean (e.g., 14 other scores such as the following: 83, 89, 78, 92, 78, 87, 93, 92, 86, 83, 90, 84, 92, 84. Now when the extreme low score of 76 is averaged in with these other 14 scores, the mean is an 86. The impact of the extreme score is lessened through the process of computing a mean, so the variation in sample means is much smaller than the variation in individual scores (that is, the range is smaller). Thus, the standard deviation is smaller with the sampling distribution than with the population distribution.

The standard deviation of the sampling distribution, $\sigma_{\bar{x}}$, is sometimes also referred to as the **standard error of the mean**. This is because deviation from the mean can be thought of as error. That is, if there is no variation (or error) every one would score exactly the same -- the mean. However, there almost always is deviation (or error) in any set of scores, which we refer to as the standard deviation. Because the sampling distribution is a distribution of sample means, the standard deviation for this distribution frequently is referred to as the standard error of the mean.

What about z-scores and percentages? As with the sample and population distributions, z-scores and percentages can be computed in exactly the same way for the sampling distribution. Let's take a sample of 15 students who scored an average of 87. What is their corresponding z-score on the sampling distribution? Well, a "raw score" (a sample mean) of 87 falls one standard deviation unit above the mean, which corresponds to a z-score of z = +1.00. Recall the general form of the z-score formula from earlier in this chapter:

$$z = \frac{(\text{whatever is on } x - \text{axis}) - (\text{mean of distribution})}{\text{standard deviation of distribution}}$$

Now let's create the z-score formula specifically for the sampling distribution. What is graphed on the x-axis are sample means, $\overline{X}$, *not* individual raw scores, X. The mean of the sampling distribution is the same as the mean for the population distribution, μ. And the standard deviation for the sampling distribution is represented by $\sigma_{\overline{x}}$. So the z-score formula for the distribution of differences looks like this:

$$z = \frac{\overline{X} - \mu}{\sigma_{\overline{x}}}$$

Using this formula we can compute the z-score corresponding with a raw score of 87 (which we already know from above is z = +1.00):

$$z = \frac{\overline{X} - \mu}{\sigma_{\overline{x}}} = \frac{87 - 85}{2} = \frac{2}{2} = +1.00$$

Now let's try a percentage problem. What percent of samples had a mean less than 87? We know that 50% of all samples have an average score below the mean of the sampling distribution ($\mu = 85$), so more than 50% of the sample must have average scores below 87 (see Figure 4.8). We also know that 68% of all sample means (raw scores) fall between +/-1 *SD* unit around the mean, or between raw scores of 83 and 87. So, 34% of all scores must fall between the mean of the sampling distribution and +1 *SD* unit (between raw scores of 85 and 87). Therefore, 50 + 34 = 84% of all sample means fall below the sample mean of 87.

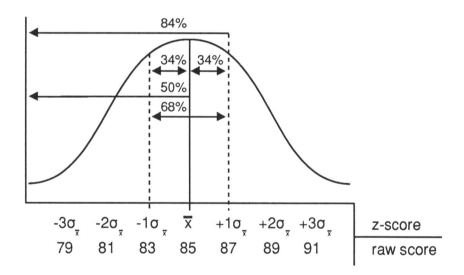

Figure 4.9 How to derive the percentage of sample means (84%) with an average score less than 87

Distribution of Differences

The final type of distribution that you will encounter is the distribution of differences. The distribution of differences is the most important distribution for hypothesis testing (which we will discuss, in the next chapter) because it allows you to compare two sample means. As with the sampling distribution, the distribution of differences does not consist of individual raw scores but rather of sample means. However, with the distribution of differences what is plotted on the x-axis is not a sample mean alone but actually the difference between pairs of sample means.

Let's consider how we created the sampling distribution. Recall that we sampled with replacement -- people drawn in any sample were replaced into the population to have the opportunity to be drawn again in another sample). We randomly sampled students (we arbitrarily chose 15 at a time), computed a sample mean, did this exhaustively (or until we got exhausted after we did this 1,000 times), and then plotted all of these sample means onto a distribution that was normally distributed. Now let's take these same sample means and instead of plotting each individual mean, let's create a difference score between two sample means. Consider two samples we drew when creating the sampling distribution:

Sample One		*Sample Two*	
Student	Test Score	Student	Test Score
1 Joseph	86	1 Marcia	81
2 Bob	79	2 Peter	75
and so forth....		and so forth...	
15 Mary	82	15 Charlie	90
$\overline{X}_1 = 82$		$\overline{X}_2 = 76$	

Let's create a difference score using these two sample means: $d = 82 - 76 = +6$. Now we will continue to do this with all of the sample means exhaustively: randomly sample 15 students, compute a sample mean, randomly sample another 15 students, compute a sample mean, subtract the two sample means creating a difference score, repeat over and over. Again, for the purposes of illustration, let's stop at what seems a reasonable point, say after we've created

500 difference scores (involving 1,000 samples). Now if we plot these 500 difference scores we obtain a normal distribution that looks like the following:

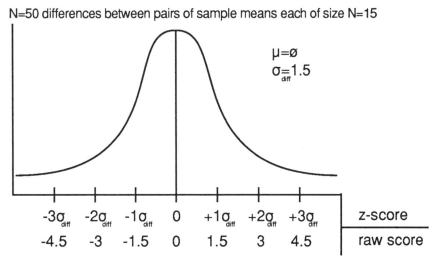

Figure 4.10 Distribution of differences with mean $\mu_{diff} = 0$ and standard deviation $\sigma_{diff} = 1.5$

This is known as the **distribution of differences**. This distribution, as with all normal distributions, has a mean, $\mu_{\bar{x}_1 - \bar{x}_2}$ (or μ_{diff}) = 0, and a standard deviation, $\sigma_{\bar{x}_1 - \bar{x}_2}$ (or σ_{diff}) = 1.5. <ins>the mean for the distribution of differences will always be equal to zero.</ins>

You may note from the graph that the mean of the distribution of differences is equal to zero. Is this by chance? The answer is no. In fact, the mean for the distribution of differences will *always* be equal to zero. That is, if you take any two randomly selected samples, compute means and subtract them, on average, all of these difference scores will be equal to zero. There is no difference in the average score of an infinite number of random samples selected from the same population. Let's see how this would work using only eight samples (i.e. four difference scores).

Using the population of 1,500 students, we know that the mean of this population is $\mu = 82$. Let's take two randomly selected samples:

74 WORKING WITH DISTRIBUTIONS

Sample One		Sample Two	
Student	Test Score	Student	Test Score
1 Joseph	86	1 Marcia	81
2 Bob	79	2 Peter	75
and so forth....		and so forth...	
15 Mary	82	15 Charlie	90
$\overline{X}_1 =$	82	$\overline{X}_2 =$	76

The difference score is $d = 82 - 76 = +6$. Now let's look at two more randomly selected samples:

Sample Three		Sample Four	
Student	Test Score	Student	Test Score
1 Ellen	86	1 Martin	81
2 Chris	79	2 Maria	75
and so forth....		and so forth...	
15 Andy	82	15 Ken	90
$\overline{X}_3 =$	*81*	$\overline{X}_4 =$	*84*

The difference score is $d = 81 - 84 = -3$. And at two more random samples:

Sample Five		Sample Six	
Student	Test Score	Student	Test Score
1 Barry	86	1 Max	81
2 Sonny	79	2 Bob	75
and so forth....		and so forth...	
15 Alejandro	82	15 Kelly	90
$\overline{X}_5 =$	*84*	$\overline{X}_6 =$	*83*

The difference score is $d = 84 - 83 = +1$. And the final two random samples for this illustration:

Sample Seven		Sample Eight	
Student	*Test Score*	*Student*	*Test Score*
1 Wally	86	1 Susan	81
2 Byron	79	2 Terry	75
and so forth....		and so forth...	
15 Adam	82	15 Alberto	90
$\overline{X}_7 =$	*81*	$\overline{X}_8 =$	*85*

The difference score is $d = 81 - 85 = -4$.

If we look at all 8 sample means and compute an average we obtain the following:

$$\mu = \frac{82 + 76 + 81 + 84 + 84 + 83 + 81 + 85}{8} = 82$$

This makes sense since we know that the mean of the population is 82. We should logically assume that if we take random samples forever, compute means, and then average the means, we should get the population mean (remember: that's why the mean of the population distribution and the mean of the sampling distribution will always be the same!). Although we only illustrated this using eight samples, this would actually be done for an infinite number of samples.

Now let's look at the average difference among sample means. The difference between samples one and two was +6, between samples three and four the difference was -3, between samples five and six the difference was +1,

76 WORKING WITH DISTRIBUTIONS

and between samples seven and eight the difference was -4. The average of these four difference scores is as follows:

$$\mu_{diff} = \frac{(+6)+(-3)+(+1)+(-4)}{4} = 0$$

So the average difference between randomly selected samples drawn from the same population is zero. That is, the mean of the distribution of differences will always be equal to zero.

As with the sampling distribution, the standard deviation for the distribution of differences, μ_{diff}, also has a special name, the **standard error of the difference** (or you may hear this referred to as the standard error of the mean difference, since it is a measure of the variance or error for the difference between sample means). As with the standard error of the mean (the standard deviation for the sampling distribution), the standard error of the difference is simply another way of describing how much scores (in this case, difference scores) deviate from the mean.

For the distribution of differences, z-scores and percentages function in exactly the same way as with the prior three normal distributions. For z-scores, let's refer back to the standard formula again:

$$z = \frac{(whatever\ is\ on\ x-axis) - (mean\ of\ distribution)}{standard\ deviation\ of\ distribution}$$

For the distribution of differences, what is on the x-axis is neither an individual raw score, X, nor a sample mean, $\overline{X}$. Instead, this is a frequency distribution of difference scores, $\overline{X}_1 - \overline{X}_2$, so this is what is graphed on the x-axis. The mean of the distribution of differences is $\mu_{\overline{X}_1 - \overline{X}_2}$ (or μ_{diff}) and the standard deviation is $\sigma_{\overline{X}_1 - \overline{X}_2}$ (or σ_{diff}). So, the z-score formula applied to the distribution of differences is the following:

$$z = \frac{(\overline{X}_1 - \overline{X}_2) - \mu_{\overline{X}_1 - \overline{X}_2}}{\sigma_{\overline{X}_1 - \overline{X}_2}}$$

or

$$z = \frac{(\overline{X}_1 - \overline{X}_2) - \mu_{diff}}{\sigma_{diff}}$$

Because the mean of the distribution of differences is always equal to zero, this formula is sometimes shortened in the following way:

$$z = \frac{(\overline{X}_1 - \overline{X}_2) - 0}{\sigma_{diff}}$$

or

$$z = \frac{(\overline{X}_1 - \overline{X}_2)}{\sigma_{diff}}$$

However, to keep clear in your mind that this really is exactly the same z-score formula as you used with the other three distributions, we suggest that you remember that there is a mean equal to zero that you are subtracting from the numerator, and you may want to use the first formula initially to familiarize yourself with this idea.

Let's try a z-score problem with the distribution of differences. Suppose you find a difference between sample means of $d = -4.5$. What z-score would this correspond to? Well, since a difference of -4.5 is three standard deviation units below the mean (see Figure 4.8; note that the standard error of the difference is 1.5), the corresponding z-score would be $z = -3.00$.

Let's also try a percentage problem. What percentage of differences between pairs of sample means are larger than +/-3? Let's review what the 68-95-99 rule would look like as applied to the distribution of differences:

78 WORKING WITH DISTRIBUTIONS

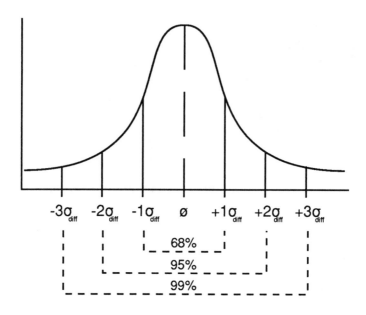

Figure 4.11 The 68-95-99 rule as applied to the distribution of differences

Now let's use this information to determine what percentage of difference between pairs of sample means fall beyond +3 and -3. A raw difference score of $d = +3$ corresponds to a z-score of $z = +2.00$ and a raw score difference of $d = -3$ corresponds to a z-score of $z = -2.00$. We know using the 68-95-99 rule that 95% of all differences will be between $z = -2.00$ and $z = +2.00$ (see Figure 4.11). So, 5% of all differences fall outside of this 95% range (2.5% of differences will be above $z = +2.00$ and 2.5% of differences will be below $z = -2.00$). Remember to follow this example working from the bottom of the graph upwards:

WORKING WITH DISTRIBUTIONS **79**

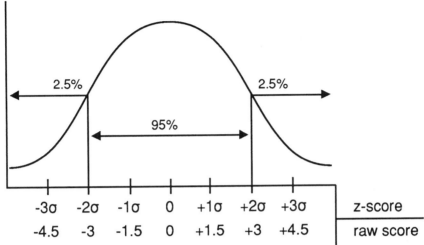

Figure 4.12 How to derive the percentage of differences (5%) beyond +/-3

Again notice how computing z-scores and percentages is exactly the same for all four of the normal distributions we discussed in this chapter! This distribution of differences will be particularly critical in the next chapter when we compare sample means to each other.

Chapter 4 Problems

1. What type of distribution represents for all of the individual scores of a *sample*?
   a. What is the symbol for the mean of these data?
   b. What is the symbol for the standard deviation of these data?
   c. What is the formula to compute a z-score for raw scores from this distribution?

2. What type of distribution represents for all of the individual scores of a *population*?
   a. What is the symbol for the mean of these data?
   b. What is the symbol for the standard deviation of these data?
   c. What is the formula to compute a z-score for raw scores from this distribution?

3. What type of distribution represents sample means computed from an infinite number of samples drawn from the same population?
   a. What is the symbol for the mean of these data?
   b. What is the symbol for the standard deviation of these data?
   c. What is the formula to compute a z-score for raw scores from this distribution?

4. What type of distribution represents differences between *pairs* of sample means that are computed from an infinite number of samples drawn from the same population?
   a. What is the symbol for the mean of these data?
   b. What is the symbol for the standard deviation of these data?

c. What is the formula to compute a z-score for raw scores from this distribution?

5. The term *statistic* is used to refer to means and standard deviations for a _____, whereas the term *parameter* is used to refer to means and standard deviations for a _____ .

6. Which of the four types of distributions (*sample, population, sampling,* or *differences*) use statistics? Which use parameters? Why do some distributions use statistics and others use parameters?

7. M and μ are symbols used to stand for the _____ as applied to different types of normal distributions.

8. SD, σ, and σ_{diff} are symbols used to stand for the _____ as applied to difference types of normal distributions.

9. What is another name for the standard deviation of the sampling distribution?

10. What is another name for the standard deviation of the distribution of differences?

11. Consider the following 9 pairs of sample means:
    11 and 3   9 and 13   11 and 11
    8 and 12   8 and 16   14 and 10
    16 and 12   5 and 5   9 and 9

    a. Using the pairs given above, compute a difference score for each pair and draw a graph of these difference scores (use the first

82 WORKING WITH DISTRIBUTIONS

two samples as a pair, the second two samples as a pair, and so on until you exhaust all pairs of sample means).

    b.   What is the name of this distribution?
    c.   What is the mean of this distribution?
    d.   What is another name for the standard deviation of this distribution?

12. A z-score is the deviation of a score from the mean as expressed in _____ units.

13. For each of the following raw scores, indicate what the corresponding z-score would be:

    a.   a raw score that is $1\frac{1}{2}$ standard deviations above the mean
    b.   a raw score that is 2 standard deviations below the mean
    c.   a raw score that is equal to the mean
    d.   a raw score that is $\frac{2}{3}$ standard deviation below the mean
    e.   a raw score that is $2\frac{3}{4}$ standard deviations above the mean

14. For each of the following sets of data (represented by the mean and standard deviation below), identify the end-points for six standard deviation units (that is between +/-3 standard deviations around the mean -- the range). It is probably helpful for you to get into the habit of *drawing* these distributions now!

    a. $\overline{X} = 15$ $SD = 3$
    b. $\overline{X} = 1,000$ $SD = 250$
    c. $\overline{X} = 75$ $SD = 5$
    d. $\overline{X} = 150$ $SD = 15$

15. A test is administered to students and the resulting scores are normally distributed. The mean of this test is 75 and the standard deviation is 8.
   a. Draw the graph for the distribution of these data and label it both in terms of raw scores and z-scores (standard deviation units).
      b. Identify the z-score for each of the following raw scores on this distribution:
   i. 83
   ii. 63
   iii. 93
   iv. 75
   v. 60
   c. Identify the percentage of students who scored in the following ranges of raw scores:
   i. between 67 and 83
   ii. between 51 and 99
   iii. above 75
   iv. above 83
   v. below 67
   vi. below 91
   vii. between 59 and 91
   viii. below 99
   d. Identify the appropriate raw score corresponding to each of the following z-scores:
   i. $z = +1.50$
   ii. $z = 0.00$
   iii. $z = -3.00$
   iv. $z = -0.25$
   v. $z = +2.33$
   vi. $z = -1.75$

16. Suppose you have a distribution of differences with a standard error of the difference, $\sigma_{diff} = 3$.
    a. Draw the normal distribution corresponding to this graph and label the raw difference scores corresponding to +/-1 SD unit, +/-2 SD units, and +/-3 SD units.
    b. What raw score difference falls at $+1\sigma_{diff}$?
    c. What raw score difference falls at $-2\sigma_{diff}$?

17. Suppose a researcher is interested in knowing about the distribution of weights of football players in the National Football league. Let's assume (yes, pro football fans, this is a somewhat gross assumption) that there are 100 players per team in the NFL and there are 28 teams.
    a. What is the population of interest to this researcher?
    b. What are the appropriate symbols for the mean and standard deviation of this population?
    c. Suppose the mean weight for the population of NFL players is 210 pounds with a standard deviation of 15 pounds. One Houston Oiler player, Tom, weighs 1.5 standard deviations above the mean NFL weight. How much does Tom weigh?

   Let's further suppose for simplicity's sake that every one of these 28 teams can be considered to be a sample.
    d. What type of distribution can be created for the weights of players on the Dallas Cowboys' team?
    e. What are the appropriate symbols for the mean and standard deviation of this distribution?
    f. If the mean weight of players on the Dallas Cowboys' team is 208 pounds with a standard deviation of 10 pounds, how many standard deviations away from the mean is Dallas player Bob if he weighs 213 pounds?

g. What percentage of Dallas Cowboys weigh between 198 and 218 pounds?

h. What percentage of Dallas Cowboys weigh more than 228 pounds?

i. If a sampling distribution were to be created using each NFL team as a sample, how many samples would be used to create the sampling distribution?

Suppose that a distribution using teams as samples is created and the mean of this resulting distribution is 200 pounds with a standard deviation of 5 pounds.

j. What is this distribution called?

k. How many standard deviations away from the NFL weight mean is the Pittsburgh Steelers' team which has an average weight of 215 pounds?

l. What percentage of NFL teams (samples) weigh more than what the Pittsburgh Steelers' team weighs on average?

CHAPTER 5
HYPOTHESIS TESTING & THE Z-TEST

What is a Hypothesis?

In this chapter we want to introduce you to the notion of hypothesis testing, which is based on the distribution of differences that we discussed in the last chapter. Arguably, we believe this is the most important chapter in the book as it ties together the concepts from the previous chapters and provides the underlying logic for all the statistical analyses in the forthcoming chapters. Hence, be sure you understand all of the concepts presented in this chapter before you move on.

We will begin by examining the term hypothesis. A **hypothesis** is a testable prediction about the relationship among variables. You formulate and test hypotheses as a regular part of your daily lives. For example, consider the statistics class you are in right now. Suppose your instructor hands out a review sheet before the test. Should you use this review sheet to study from for the exam or not? You and your friends may hypothesize that students who study from a review sheet will score differently, on average, than will students who do not study from the review sheet. You don't know whether the review sheet will help students score higher or whether it will actually hurt their test performance, so you simply hypothesize it will make some difference in their exam scores. You decide to perform a mini-experiment to test this idea. You put all your friends' names into a hat (say, 10 of them) and randomly pick out five names to be those who will use the review sheet to study for the midterm exam. The other five of your friends are randomly assigned not to use the review sheet. After the midterm you compute the average grade for your five friends who studied from the review sheet and for your five friends who did not use the review sheet. Your friends who studied from the review sheet scored an average of 87 and your friends who did not study from the review sheet scored an average of 78. Based on these data, you might conclude that your hypothesis was confirmed: those who studied from the review sheet did indeed score differently than those who did not study from the review sheet (in this case, the review sheet actually seemed to help test performance).

Although you may not approach problems like this quite so systematically, hypothesis testing is a direct extension of ways you normally learn and interact with the world. In a similar fashion, researchers formulate hypotheses based on reviewing what they already know and test those hypotheses by conducting research studies. For example, a psychologist interested in the issue of learning environments might hypothesize that recall of information is greater if it occurs in the same location in which the material was initially learned than if the recall occurs in a different location. A communication researcher examining the topic of pornography in society might hypothesize that men who are exposed to violent pornography will behave more aggressively toward women than will men who are exposed to nonviolent pornography. A sociologist might hypothesize that women who are raised in families in which the mother worked outside of the home will be more likely to work outside the home themselves than will women raised in families in which the mother worked at home. An economist might hypothesize that there will be a difference in recovery rates following a recession in primarily agricultural states compared to primarily technological states.

Forming Hypotheses

Many hypotheses that researchers formulate can be tested by collecting data from experiments; other hypotheses are tested using data collected from survey research. If you recall from Chapter 2, data can be collected at four different levels: nominal, ordinal, interval, and ratio. All of these types of data can be used for hypothesis testing. What we will be covering in the remainder of this book are different statistical tests that allow you to test hypotheses based on the type of data you have collected and the number of independent variables you have.

Let's take a look at the hypothesis that you formulated involving the statistics review sheet: students who study from the review sheet will score differently than will students who do not study from the review sheet. We can call this the research hypothesis because it is what you as the researcher are predicting or hypothesizing. Often the research hypothesis is referred to as the **alternative hypothesis**, and is symbolized H_a. But alternative to what? It is an alternative to the **null hypothesis**, (symbolized H_o), which is the hypothesis that is directly tested. What you are hypothesizing as the researcher is alternative to the idea that students who study from the review sheet will score the same as students who do not study from the review sheet. This is called the null hypothesis because it is essentially predicting no difference. Regardless of what type of data you have, the basic principle of forming hypotheses is the same: you will have a research or alternative hypothesis, which is what the researcher predicts, and the opposite of the researcher's prediction, the hypothesis of no difference or the null hypothesis.

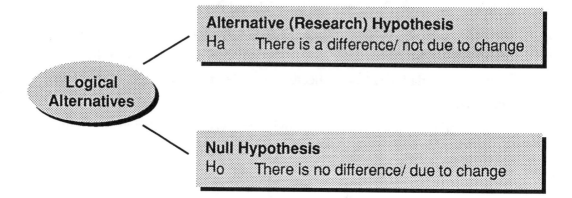

Figure 5.1 The relationship between the alternative (research) and null hypotheses: logical alternatives

Let's consider how we might illustrate these hypotheses symbolically rather than in words. First, consider what the dependent measure is. Remember: the dependent variable is the variable on which you measure an outcome; this is the variable that you will analyze statistically. In this example, the outcome or dependent variable is test score. Since test score ranges from 0 to 100 with one point serving as an equal interval and zero being an absolute zero point (you can't score less than nothing on a test hopefully), this is data measured at the ratio level. In the next few chapters all of the statistical tests we will be describing apply to both interval- and ratio-level data. When looking at interval- or ratio-level data we know that we can compute a mode, median, and mean (from Chapter 3). In this example, we are looking at mean test scores; that is, the mean difference between those who studied from the review sheet and those who did not. Consider again the research (alternative) hypothesis: students who study from a review sheet will score differently (on average, using the mean) than will those who do not study from a review sheet. In this example we have two groups (or conditions or levels): those who study from a review sheet versus those who do not study from a review sheet. So, we are conducting this experiment with one independent variable, presence of studying from review sheet, which has the two levels described above. Now consider a mean for an entire population that consists of those who study from a review sheet, μ_{RevSh}. And, consider a mean for a different population, one that does not study from a review sheet, $\mu_{No\ RevSh}$. The research or alternative hypothesis looks like the following:

$$H_a: \quad \mu_{RevSh} \neq \mu_{No\ RevSh}$$

The corresponding null hypothesis, or hypothesis of no difference, looks like the following:

$$H_0: \quad \mu_{RevSh} = \mu_{No\ RevSh}$$

That is, students who study from the review sheet will score no differently than students who do not study from the review sheet. The null hypothesis will always be *opposite* from what the researcher predicted. Notice that the null and the research hypotheses use the mean for the population, μ and not the mean for the sample, $\overline{X}$.

The research hypothesis presented above is a **non-directional** or **two-tail** hypothesis. It is non-directional because there is no specification about which direction the means of the two groups will fall in, simply that they will be *different* from one another. We will see why this form of a hypothesis is called two-tail in a moment.

As a savvy student you might have formulated a different research hypothesis instead of the one above. You might hypothesize that the review sheet actually will *help* students perform better on the midterm exam. So, your research hypothesis would be: students who study from a review sheet will score higher than will those who do not study from a review sheet. In this instance the research or alternative hypothesis would look like the following:

$$H_a: \quad \mu_{RevSh} > \mu_{No\ RevSh}$$

That is, the mean score of those who studied from the review sheet will be higher than the mean score of those who did not study from the review sheet. Since the null hypothesis is always the opposite of what the researcher predicted, the corresponding null hypothesis would look like the following:

$$H_0: \quad \mu_{RevSh} \leq \mu_{No\ RevSh}$$

Those who studied from the review sheet will score no differently from (or less than) those who did not study from the review sheet.

This form of the research hypothesis is referred to as a **directional** or **one-tail** hypothesis. It is called directional because the researcher is specifying which *direction* the pattern of means will fall in. This can be either greater than (>) or less than (<) depending upon the problem. Again, we will discuss below why this form of a hypothesis is called one-tail.

Not all hypotheses utilize means. Depending upon the type of data you have and what you are interested in testing, your hypotheses might involved modes (that is frequency of occurrence) or medians, or it might even involve something totally different from any of the three measures of central tendency (such as correlations, which we will present in Chapter 9). Consider an example from the medical field. Suppose that a new flu shot comes out that is supposed to prevent you from catching one of a whole strain of colds and flus. We might conduct a study to see whether the shot really is effective in preventing colds. Half of a group of participants are randomly assigned to receive the shot and the other half do not receive the shot. We then expose all of the participants to a bunch of cold and flu viruses and see how many of them get sick. The research hypothesis is that a lower number of people who got the flu shot will get sick than will those who did not get the flu shot. The null hypothesis is that the same number of people (or more people) will get sick among those who got the flu shot as compared to those who did not. Although what we have described here is actually a chi-square problem (Chapter 11), the basic principle of forming hypotheses is the same. In this case, we cannot compute means on the dependent measure because we do not have interval- or ratio-level data. We actually have nominal-level data: the number of people who got sick. In this case we use this frequency of occurrence to statistically test (using the chi-square) whether fewer people got sick among those who had the shot than did not. In sum, we have three main alternative hypothesis than can be posed by the researcher:

Questions the Researcher Asks		Labeled	Test Used
In Words	Symbolically Represented		
1. Is there a difference (H$_o$: There is no difference)	H$_a$: $\mu_1 \neq \mu_2$ H$_o$: $\mu_1 = \mu_2$	Nondirectional Hypothesis	Two-Tail Test
2. Is x greater than y (H$_o$: x is less than or equal to y)	H$_a$: $\mu_1 > \mu_2$ H$_o$: $\mu_1 \leq \mu_2$	Directional Hypothesis	One-Tail Test
3. Is x less than y (H$_o$: x is greater than or equal to y)	H$_a$: $\mu_1 < \mu_2$ H$_o$: $\mu_1 \geq \mu_2$	Directional Hypothesis	One-Tail Test

Table 5.1 Alternative (research) and null hypotheses

Probability

With hypothesis testing there are two hypotheses (i.e., two possibilities): 1) that the researcher's hypothesis (the alternative hypothesis) is correct (e.g., people who used the review sheet scored differently than people who did not use the review sheet), or 2) that the null hypothesis is correct (e.g., that people who used the review sheet did not score any differently than people who did not use the review sheet). So how do we know which one is correct? Well, we assume that the researcher's hypothesis is wrong and that the default is the null hypothesis. We need to be convinced that there really is a difference. This is sort of like being assumed innocent until proven guilty. There is assumed to be no difference until the data are statistically proven (with some chance of error) to show a difference. To be convinced that we should accept the researcher's prediction and reject the null hypothesis, we must be very, very certain that the null hypothesis is wrong. In fact, researchers have decided by convention that we must be at least 95% certain that the null hypothesis is wrong before we can accept the researcher's prediction. In other words, if we accept the researcher's hypothesis and reject the null hypothesis there must be a probability of less than 5% (notated as $p < .05$) that we are wrong in making this decision.

Let's consider how this notion of hypothesis testing and probability might work in the example of the flu shot described earlier. Suppose the new shot costs $200. How confident would you want to be that this new shot actually does prevent colds and flus before you spend the money to receive it? 50% confident? 75% confident? 95% confident? 99% confident? Presumably most people would want to be 95% confident (or more) that the flu shot actually works, on average, before they spend the money on it (not to mention the time involved in going to the doctor and the pain of actually getting the shot). So you want the probability to be $p < .05$ that the flu shot does not work. That is, you want to reject the null hypothesis at the $p < .05$ level (so that there is no more than a 5% chance that there is no difference in the number of people who caught colds between those who got the shot and those who did not). There is a 95% chance that the research hypothesis is the correct one: that those who got the flu shot were *significantly* less likely to catch a cold than were those who did not get the flu shot.

The term **significant difference** is used to refer to a difference that is not due to chance. For instance, let's consider again the example with the students who got a review sheet versus those who did not. Suppose the group who got a review sheet scored an 84 and those who did not get the review sheet scored an 83. Did the review sheet really and truly improve test scores among the group who had it, or is this difference (84 versus 83) purely a random difference among two different groups? Probably the latter.

Consider another example. We randomly divide your statistics class in half and compute the grade point average for the right and left halves of the room. The left half of the room has a GPA of 3.14 and the right half of the room has a GPA of 3.12. Is the right half of the room really, truly smarter (or studies more) than the left half? Probably not. This difference is most likely due purely to random chance because of how people happened to sit down today. But suppose we found the right half of the room had a GPA of 3.6 and the left half of the room had a GPA of 2.4. In this case we might be more convinced that this is a significant (true) difference. There is something truly different about the right and left halves of the room (e.g., all the heavy partiers decided to sit on the left half of the room together and all the Phi Beta Kappa students decided to sit on the right half of the room). The right and the left halves of the room truly represent two different groups or populations with respect to GPA. So how do you know, then, when you have a real, true, significant difference between groups, and when a difference is simply due to error or chance? This is the fundamental notion of hypothesis testing. How different do two (or more!) means (or frequencies, or medians, or whatever) need to be from one another in order to reject the null hypothesis and conclude that the difference is significant (and the researcher's hypothesis should be accepted)?

Hypothesis Testing

Let's see how hypothesis testing would work with two groups (e.g., a group that got a review sheet and a group that did not) and a dependent variable with interval- or ratio level data (e.g., midterm exam score). Let's refer back to the distribution of differences we created in Chapter 4. Recall that this frequency distribution displayed a difference score between pairs of sample means, which we created by exhaustively sampling (with replacement) two samples of some random size (we picked 15 for each sample) from the entire population (of size 1,500 students in the example), computing a mean for each sample, subtracting the two means from one another, and plotting each difference score with corresponding frequency. This resulting distribution represents a distribution of differences for statistics students in an introductory statistics class:

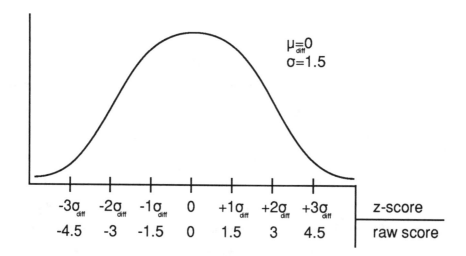

Figure 5.2 Distribution of differences with mean ($\mu = 0$) and standard deviation ($\sigma_{diff} = 1.5$)

Suppose we show up one day with 15 scores of some people (we don't tell you where we got these people from) and have them take your midterm statistics exam. We find an average for them: $\overline{X} = 100$!! Then we randomly sample 15 students out of the 1,500 in the entire population of introductory statistics students. Their mean is $\overline{X} = 82$. Would you conclude that the 15 people we showed up with (who got a mean of 100 on the exam) are 15 of the 1,500 students in the population of introductory statistics students? Probably not. In fact, these people later tell you that they are 15 professors who teach statistics, so it is not surprising that they averaged a perfect score on an introductory statistics exam. Would you say these 15 statistics professors were from the same population as the 1,500 introductory statistics students? Of course they are not.

If you recall, the mean difference on a distribution of differences is always equal to zero (on average, there is no difference between sample means from samples that are all drawn from the same population), and there is some standard deviation. In the sample problem, the standard error of the difference is $\sigma_{diff} = 1.5$ (see Figure 5.2). What percentage of differences among sample means in the population of introductory statistics students are larger than 4.5 (either positively or negatively, that is, whether the difference is +4.5 or -4.5?). Recall the 68-95-99 rule: 68% of all differences (since we're working with the distribution of differences) fall within +/-1 SD unit around the mean (between raw score difference of +/-1.5 in this example); 95% of all differences fall within +/-2 SD units around the mean (between raw score differences of +/-3), and 99% of all differences fall within +/-3 SD units around the mean (between raw score differences of +/-4.5). So, if 99% of all scores fall within +/-4.5 raw score difference points around the mean, then only 1% of all difference scores must be larger (in absolute value terms, that is, either + or -) than 4.5.

Now let's look at the difference between the sample of 15 mysterious people we showed up with who had a mean of $\overline{X} = 100$ and the 15 randomly selected introductory statistics students with a mean of $\overline{X} = 82$. The difference here is $d = 100 - 82 = 18$! Wow! This is way outside of that +/-4.5 range. This difference, in fact, is so large that less than 1% of all difference scores computed from samples drawn from the population of introductory statistics students would ever be this large. Perhaps a better conclusion is that the sample we showed up with who scored a 100 on average are really not from the population of introductory statistics students but rather, are from some other population, such as statistics professors. Based on the distribution of differences, we can be more than 99% confident that this is the case; that the two sample means are *significantly different* from one another. Now remember that there is the possibility that we actually did pull this sample from the population of 1,500 statistics students, since there are a tiny fraction (less than 1%) of all differences between sample means that could be +/-4.5 or larger. However, it is so improbable that this is the case ($p < .01$) that a much better conclusion is that these two samples were drawn from different populations: a population of introductory statistics students and a population of statistics professors. And, we are more than 99% confident that this is the case in this example.

Now let's approach a similar problem related to the review sheet. Let's take the population of 1,500 introductory statistics students and draw a random sample of 30 students. Let's then take these 30 students and randomly assign them to either use the review sheet to study for the midterm exam or to not use the review sheet. (Remember from chapter 2 that random sampling and random assignment are two different things. Random sampling is a method for gathering a sample in which everyone in the population has an equal chance of being included in the final sample. In the case of drawing the sample of size 30, we might have put all 1,500 names of students in the population into a hat and pulled out 30 -- this is random sampling. Notably, however, almost all experiments do *not* use random samples. Random assignment refers to the process by which people are placed into groups or conditions in an experiment once you already have selected the sample by whatever means, randomly or not. Using the 30 sampled students, we might flip a coin for each one; if it is heads, we randomly assign the student to study from the review sheet and if it is tails we randomly assign the student to not study from the review sheet. Random assignment is used to place people into groups in an experiment regardless of how you selected your sample. Random assignment is important because it insures that any initial differences among those selected in the sample, such as GPA, interest in school, amount of partying, or any other characteristic you think might influence the results is equally distributed across conditions. Thus, any difference observed on the dependent measure, such as exam score, cannot be explained by differences in any variable other than the one you manipulated (e.g. studying from the review sheet).

After the midterm exam we compute a mean score for the 15 students who were randomly assigned to study from the review sheet: they scored an average of 86.5. We also compute a mean for the 15 students who did not study from the review sheet: they scored an average of 83. So, did the review sheet *cause* students to score differently than students who did not use the review sheet? That is, is this a statistically significant difference or is this a difference that can be explained by how people were randomly assigned to conditions (i.e. chance)? To answer this question let's refer back to the distribution of differences. The mean difference between the two samples is $d = 86.5 - 83 = 3.5$. So, how likely is it that these two samples still represent the same population (a population where studying from a review sheet is irrelevant to exam score)? We can compute a z-score using the raw score difference of 3.5 and the standard error of the difference $\sigma_{diff} = 1.5$. Since $d = +3.5$ falls 2.33 *SD* units above the mean, we know that the corresponding z-score is $z = +2.66$. So, what percentage of scores fall above a z-score of +2.33? Well, since 95% of all scores fall within +/-2 *SD* units, we know that less than 5% of all difference scores fall outside of this range. So, there is a $p < .05$ chance that these two samples actually represent the same population, and a 95% chance that the sample that got the review sheet is now representative of a different population: a population that knows more about statistics because they studied from a review sheet. So, the best conclusion is to reject the null hypothesis (the hypothesis that there is no difference in midterm exam scores between those who studied from the review sheet and those who did not) at the $p < .05$ level (since +2.33 is outside of the +/-2 *SD* unit range). We are 95% sure that studying from the review sheet *caused* students to score higher on average than those who did not study from the review sheet: we reject the null and accept the alternative hypothesis. Ultimately, a researcher hopes to reject the null hypothesis (that there is no difference) and accept the alternative hypothesis (that there is some difference) because s/he is predicting some difference in the alternative (research) hypothesis. For example, presumably your statistics professor designed the review sheet with the intent of helping students learn the course material better (and score higher on the exam). Therefore, the goal is to reject the idea that the review sheet doesn't help (the null hypothesis).

At this point we should probably mention a couple minor technicalities that we have overlooked. First, the distribution of differences is sometimes referred to as the **z-distribution** because z-scores are computed from it (which we can do for the other types of distributions -- sample, population, sampling-- as well). We will see in the next chapter that the z-distribution (which is normally distributed) is simply a theoretical version of a whole class of distributions (which are not quite normally distributed) known as t-distributions. For now, you can think of the z-distribution and the distribution of differences as exactly the same thing.

Second, what we just conducted using the z-distribution (with the example of the professors' scores on the midterm exam versus introductory statistics students, and with the example of students who received the review sheet versus those who did not) was your first statistical test -- the z-test! That wasn't so hard, was it? You probably did it without even realizing you were actually conducting a statistical test.

Third, the numbers we have been using for the 68-95-99 rule have only been approximations. With the 68-95-99 rule we noted that 68% of all scores fall within +/- 1 *SD* unit around the mean, 95% of all scores fall within +/-2 *SD* units around the mean, and 99% of all scores fall within +/-3 *SD* units around the mean. We used this piece of information to determine that a mean score difference of 3 in the example above, which corresponded to a z-score of +/- 2.00, fell outside of the 95% range. Well, technically, it is not +/-2 *SD* units, that include 95% of the distribution and excludes the extreme outermost 5%.

Technically, the z-score value is +/- 1.96 *SD* units. Similarly, it is not +/-3 *SD* units that include 99% of the distribution and exclude the extreme outermost 1%. Technically, the z-score value is +/-2.58 *SD* units. This is just a technicality for now, but those numbers will become more important shortly as we discuss one-tail and two-tail hypothesis testing and the concept of the t-test in the next chapter.

One-Tail and Two-Tail Hypothesis Testing

Now that you have conducted your first statistical test, the z-test, let's review what we did. First, recall what the 68-95-99 rule looks like for the distribution of differences:

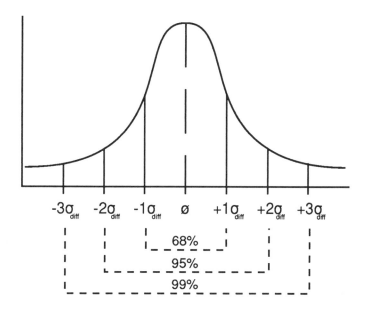

Figure 5.3 The 68-95-99 rule for the distribution of differences

Using the example of students who studied from the review sheet versus those who did not, the alternative (research) hypothesis stated in words is:

> H_a: Students who study from a review sheet will score differently on the midterm exam than will students who do not study from a review sheet.

The corresponding null hypothesis stated in words is:

> H_0: Students who study from a review sheet will score no differently on the midterm exam than will students who do not study from a review sheet.

Stating these two hypotheses symbolically we obtain the following:

> H_a: $\mu_{RevSh} \neq \mu_{No\ RevSh}$
> H_0: $\mu_{RevSh} = \mu_{No\ RevSh}$

The dependent measure is students' score on the midterm exam (which is ratio-level data), so we are using means here. The independent variable is presence of review sheet with two levels (or groups): present, absent. And the researcher's goal is to reject the null hypothesis and accept the alternative hypothesis.

The mean for students who got the review sheet was $\overline{X} = 86$, and the mean for students who did not get the review sheet was $\overline{X} = 83$. The standard deviation for the distribution of differences in this example was $\sigma_{diff} = 1.5$ (see Figure 5.2). The z-score value is $z = +2.33$.

To test the null hypothesis we decided that if fewer than 5% of all differences between sample means from the same population were larger than +/-1.96 (which excludes the extreme outermost 5% of the distribution), then we reject H₀ and conclude that the two groups are significantly different from one another. So, anything that falls within +/-1.96 *SD* units (approximately +/-2.00 according to the 68-95-99 rule) is acceptable: we do not reject (accept) H_0, since 95% of all differences fall within this range. Anything larger than +/-1.96 is not acceptable: we reject H_0, since only 5% of all differences fall outside of this range (where the samples actually are representative of the same population). We can illustrate this idea as follows:

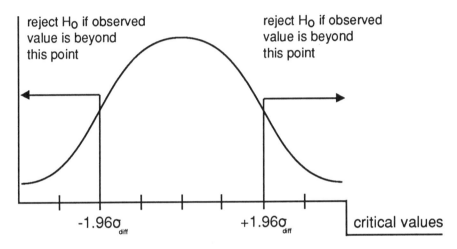

Figure 5.4 Rejecting the null hypothesis using a two-tail, non-directional hypothesis test at the $p < .05$ level

Remember, that only the 5% most extreme differences fall beyond +/-1.96 sdiff (we had been approximating to +/-2.33 σ_{diff} with the 68-95-99 rule). The probability is less than .05 ($p < .05$) of getting such extreme differences between means and having the two groups still represent the same population when such a large z-score value is obtained (larger than +/-1.96). Hence, there is only a 5% chance that these two groups represent the same population and a

95% chance that these two groups represent different populations. Thus, the decision rules for rejecting or accepting the null hypothesis can be stated simply as follows:

> Reject H_0 – z observed > z critical
>
> Accept H_0 – z observed < z critical

Table 5.2 Decision rules for rejecting/accepting the null hypothesis

Let's return again to the review sheet problem. The value we calculated for the z-score (not raw score) difference between means was $z = +2.33$. This calculated value of +2.66 is larger in absolute terms (i.e., you can ignore the positive or negative sign) than the critical value (the cut-off point for rejecting H_0) of +/-1.96. Thus, we reject the null hypothesis, since 2.33 falls in the extreme outermost 5% of the distribution. We conclude that there is a significant difference in midterm exam scores between students who studied from the review sheet and those who did not. Note that we will use the terms "calculated value" and "critical value" throughout this book. The **calculated value** for any statistical test refers to the number that you calculate using the data you collected. The **critical value** refers to some cut-off value that you need to exceed in order to reject the null hypothesis.

What we conducted was a non-directional or two-tail z-test. We know it is non-directional because we did not specify in the alternative (research) hypothesis which group would score higher than which other group, but simply that the two groups would score differently. This type of test is also called two-tailed because we used both tails (the positive and negative ends) of the z-distribution for hypothesis testing (see Figure 5.5). In other words, whether the review-sheet group scored higher than the no-review-sheet group or the no-review-sheet group scored higher than the review-sheet group doesn't matter. The alternative (research) hypothesis simply predicts that the two groups' scores will differ, so we test that the difference score is either positive or negative, which means we use both tails of the z-distribution.

Suppose, for example, that the review-sheet group scored an 88 and the no-review-sheet group scored an 85. The difference here is $d = 88 - 85 = 3$. The z-score has a value $z = +2.00$. Since $+2.00$ is larger in *absolute* terms than 1.96 (outside of the 95% range in Figure 5.5), we reject the null hypothesis and conclude the two groups are significantly different from one another. But what if we had subtracted the means the other way, no-review-sheet mean minus review-sheet mean? In this case, $d = 82 - 86 = -4$, which corresponds to a z-score of $z = -2.00$. Again, this z-score value is larger in absolute terms than 1.96 (the critical value), and the statistical decision is exactly the same: reject the null hypothesis. The direction in which you subtract the two means when computing a difference score does not matter for a two-tail test: the statistical decision will be exactly the same. (Although the direction of means does matter with a one-tail test, we will show you a way to conduct this test shortly where you can subtract the means in either order as well).

Let's try another two-tail z-test example. Assume the same non-directional (two-tail) research hypothesis regarding review sheets (that there will be a difference in midterm scores between those who studied from the review sheet and those who did not), and the following sample means:

$$\overline{X}_{RevSh} = 72$$
$$\overline{X}_{No\ RevSh} = 81$$

Is there a statistically significant difference between these two groups or is the observed 9-point difference on the midterm exam simply due to random chance? First, because we are not predicting the direction of difference between the means, we are using both tails of the z-distribution. Now let's compute a z-score difference for these two groups, and this time let's use the z-score formula to do this:

$$z = \frac{(\overline{X}_1 - \overline{X}_2) - 0}{\sigma_{diff}}$$

or

$$z = \frac{(\overline{X}_1 - \overline{X}_2)}{\sigma_{diff}}$$

So, plugging the two sample mean values into the equation, and knowing that the standard error of the difference is $\sigma_{diff} = 1.5$ (as it was before) we get the following:

$$z = \frac{(72 - 81)}{1.5} = \frac{-9}{1.5} = -6$$

Does the calculated value $z = -6$ fall outside the 95% range (marked by z-scores of +/-1.96)? It sure does, so we can definitely reject the null hypothesis at the $p < .05$ level and conclude that students who studied from the review sheet scored significantly different than those who did not. But, we still have a 5% chance of being wrong in this decision; that is, 5% of differences from the *same* population actually will fall beyond the 1.96 endpoints (2.5% will fall above +1.96 and 2.5% will fall below -1.96). But, since -6 is so much larger (in absolute value terms) than 1.96, we might be able to exclude an even smaller percentage of difference scores than 5%.

Again recall the 68-95-99 rule. Approximately +/-3 *SD* units include 99% of all scores (differences on the distribution of differences) and exclude only 1% of all differences. In reality, "approximately +/-3 *SD* units" is exactly

+/-2.58 *SD* units. So, if we look at this graphically for the distribution of differences we obtain the following:

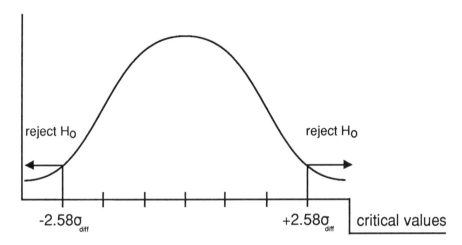

If observed value > critical value reject H_0
If observed value < critical value accept H_0

Figure 5.5 Rejecting the null hypothesis using a two-tail, non-directional hypothesis test at the *p* <.01 level

Now, where does the calculated value of $z = -6.00$ fall relative to the critical value of +/-2.58 needed to reject H_0 at the $p < .01$ level. We already know we can reject H_0 at $p < .05$; now we are trying to see if we can do any better than a 5% chance of being wrong. Since the calculated value of -6 is larger (in absolute terms) than the critical value of 2.58, we can reject the null hypothesis at the $p < .01$ level. That is, we are 99% confident that these two sample means represent two different populations (a higher scoring population because they did not study from the review sheet, and a lower scoring population because they did study from the review sheet). There is only a 1% chance that these two sample means actually represent the same population of introductory statistics students (i.e. that the review sheet did not cause students to score lower). So, we have only a 1% (rather than a 5%) chance of making an incorrect decision when we reject the null hypothesis. This is preferable if you can do it because you are more certain about your decision. Consider if we would present these results to you: we are 99% confident that students who study from a review sheet score lower on average than students who do not study from a review sheet. Would you study from the review sheet? We wouldn't!

Conventionally, researchers require at least enough confidence to reject the null hypothesis at the $p < .05$ level (95% confident) before accepting the alternative (research) hypothesis. The more confident you can be, though, the better, so you should try to reject at the $p < .01$ level if you can (and be 99% confident). Depending on the nature of the research project, researchers relax this rule or make it more stringent. For example, if the research project involved a new type of drug, researchers might want to conclude it should be given to real people only if the null hypothesis can be rejected at the $p < .001$ level, meaning there is only a 1 in 1,000 chance that the drug makes no difference. They are 99.9% confident that the drug is effective on average.

The hypotheses we have been testing so far with respect to the review sheet have been non-directional, or two-tailed, because we did not predict which group (the review-sheet group or the no-review-sheet group) would score higher or lower on the midterm exam. Presumably, in designing the review sheet, your professor was attempting to help students learn the course material better. So, in some ways, he or she had the research hypothesis that:

H_a: Students who study from the review sheet will score higher on the midterm exam than will students who do not study from the review sheet.

The following is the corresponding null hypothesis, which is the opposite of the alternative (research) hypothesis (and the one that is tested statistically):

H₀: Students who study from the review sheet will score no differently than (or lower than) students who do not study from the review sheet.

We can represent these hypotheses symbolically as follows:

H_a: $\mu_{RevSh} > \mu_{No\ RevSh}$
H_0: $\mu_{RevSh} \leq \mu_{No\ RevSh}$

Now, what we actually have here is a directional, or one-tail hypothesis. It is directional because the researcher is predicting the direction in which the means will fall. When you conduct this type of test you are using only one tail of the z-distribution (either the positive, rightmost tail, or the negative, leftmost tail), and you put all of the error (5% or 1% chance of being wrong) into that one tail. Let's look at the two possibilities for rejecting the null hypothesis at the $p < .05$ level:

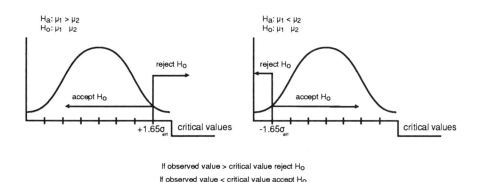

Figure 5.6 Rejecting the null hypothesis using a one-tail, directional hypothesis test at the $p < .05$ level

So, if the calculated value you obtain is larger *in the predicted direction* than the critical value, you reject the null hypothesis at the $p < .05$ level, else you cannot reject H_0. Notice for a one-tail test at the .05 level that the critical value is not +/-1.96 but rather it is 1.65 (either positive or negative depending on which graph you are using). Why would the critical value be smaller with the one-tail test than with the two-tail test? Think back to the 68-95-99 rule. What percentage of scores (difference scores for the distribution of differences) fall above a z-score of +2 (actually, +1.96)? Since we know 95% of scores fall between +/-1.96, and 5% of all scores fall beyond these values this, we know that 2.5% of all scores fall above +1.96 (see Figure 5.5). Similarly, 2.5% of all scores fall below -1.96. What percentage of scores fall above a z-score of +1.00 or -1.00? We know that 68% of all scores fall between +/- 1 *SD* around the mean, so approximately 34% fall outside of this range. Thus, 16% of all scores must fall above +1.00 *SD* and 16% must fall below -1.00 *SD*. So, using the positive tail to illustrate, if 2.5% of scores fall above +1.96 and 16% of scores fall above 1.00, then 5% of all scores must fall above some value between 1.00 and 1.96. This value happens to be 1.65 (see Figure 5.7). So, in a one-tail distribution, you need to obtain a calculated value larger than the critical value of 1.65 *in the predicted direction* (either positive or negative) in order to reject the null hypothesis.

  So, which of the two tails, positive or negative, do you use in testing the null hypothesis? Well, this depends on which direction you choose to subtract the sample means in and what the alternative (research) hypothesis is. Since we predicted that review-sheet students will score *higher* than no-review-sheet students this implies a positive difference so we should use the positive tail of the z-distribution. If we do the problem this way, we need to be sure to subtract the two sample means in the same order as stated in the hypothesis: $\overline{X}_{RevSh} - \overline{X}_{No\ RevSh}$. If we subtract the means in the other order and use the positive tail we will run into problems. Alternatively, we could correctly look at this same problem in the following way: no-review-sheet students will score *lower* than review-sheet students, which implies a negative difference. Again, we need to subtract the means in the order of this statement: $\overline{X}_{\text{No RevSh}} - \overline{X}_{\text{RevSh}}$. And here we would refer to the negative tail of the z-distribution. Be sure to use the correct tail for the direction you subtract the means, which should be done to be consistent with the alternative hypothesis.

Let's take the example from earlier in the chapter where we obtained the following means:

$$\overline{X}_{\text{RevSh}} = 88$$
$$\overline{X}_{\text{No RevSh}} = 85$$

Since the directional research hypothesis predicts that review-sheet students will score higher than no-review-sheet students we will subtract the means in this direction and use the positive tail of the z-distribution. With a standard error of the difference $\sigma_{\text{diff}} = 1.5$, we computed a z-score of $z = +2.00$ for this problem. Now, did students who used the review sheet score significantly (not due to chance) *higher* than students who did not use the review sheet? In order to reject the null hypothesis, the calculated value of +2.00 must be larger than the critical value of +1.65 (note that the sign of the calculated value *does* matter with the one-tail test), which it is. So, we reject H_0 at the $p < .05$ level and conclude that students who studied from the review sheet did score significantly higher than those who did not study from the review sheet with a 5% chance of being wrong. Can we do any better than a 5% chance of being wrong?

Well, just like the two-tail z-test can be conducted at the .05 and .01 levels, so too can the one-tail z-test. The two possibilities for a one-tail z at the $p < .01$ level look like the following:

112 HYPOTHESIS TESTING AND THE Z-TEST

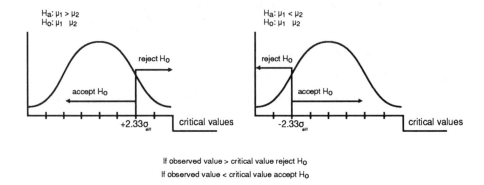

Figure 5.7 Rejecting the null hypothesis using a one-tail, directional hypothesis test at the $p < .01$ level

Again, notice two things about the critical value of 2.33 for the one-tail z. First, this value is larger than the critical value of 1.65 used for testing the one-tail z at the .05 level. This makes sense, because you need to move further out on the distribution (to a higher z-value) to include more of the distribution (moving from 95% to 99%) and exclude less of the distribution (moving from 5% outside the critical value to 1%). So the critical value to exclude the extreme outermost 1% of the z-distribution must be smaller than the critical value to exclude only the outermost 5% of the distribution. Second, the critical value of 2.33 for the one-tail z at the .01 level is smaller than the critical value of 2.58 for the two-tail z. Again, this is because we are not dividing the error (1%) between two tails (.5% in one tail and .5% in the other tail for the two-tail test), but rather we are putting all 1% of error in only one tail of the z-distribution.

Now, can we reject H_0 at the .01 significance level (we already know we can reject H_0 at $p < .05$ for this problem)? In this case, the calculated value of +2.00 is not larger than the critical value of +2.33 (remember that we are using the positive tail for this particular problem). So, we cannot reject H_0 at the $p <$

.01 level. The final conclusion, then, must be to reject H_0 at the $p < .05$ level and accept a 5% chance of being wrong.

Let's try another problem, again using the same directional research hypothesis and this time using the other set of means we obtained earlier:

$$\overline{X}_{RevSh} = 72$$
$$\overline{X}_{No\ RevSh} = 81$$

The calculated z-score value was $z = -6.00$ (notice that we subtracted the means in the order stated in the research hypothesis: review-sheet mean minus no-review-sheet mean). Did those who got the review sheet score significantly higher than those who did not use the review sheet? Well, let's start at the .05 level using the positive tail of the distribution again because the prediction in the research hypothesis indicated a positive difference in sample means. Is the calculated value of $z = -6.00$ larger than the critical value of $z = +1.65$? No. So, the conclusion is to not reject (accept) the null hypothesis. There is no difference between the scores of those who used the review sheet and those who did not.

But wait. By looking at the two means and the large (albeit negative) z-score it appears that the review sheet actually caused those who used it to perform worse!!! Isn't there any way to say this? Well, technically, no. Remember when we conducted the two-tail test we were able to test for a difference in sample means in either direction (i.e., either that the review-sheet group scored better or worse than the no-review-sheet group). However, with the one-tail test we cannot conclude a difference in the direction opposite than prediction (because we predicted the direction the means would fall in and they did not fall in this direction). So, you give something up when you conduct a one-tail test: the ability to make any statement about a difference in the opposite than predicted direction (no matter how large that difference is). Even if those who got a review sheet scored a zero on average, you would still have to not reject (accept) the null hypothesis because the research hypothesis predicted that the means would fall in the other direction (with the review-sheet group scoring higher). So why wouldn't you always want to conduct a two-tail test then, since you can examine differences between means in both directions? Well, the answer is that a one-tail test is more powerful, and we will show this below shortly.

In the meanwhile, let's look a second way to conduct the same one-tail test as above but where we do not need to worry about the direction in which the two sample means are subtracted. This method will always yield the exact same results as the way we described above for conducting a one-tail test, but if you're a little confused about subtracting the means in the same direction as predicted by the hypothesis and whether to use the positive or negative tail when conducting the one-tail test, then you may prefer the following two-step method.

For this second method, start by examining the two sample means. Do the means fall in the direction predicted by the research hypothesis? The directional hypothesis predicted that the review-sheet group would score higher than the no-review-sheet group. So when looking at a pair of sample means, the review-sheet group mean *must* be higher than the no-review-sheet group mean, else we automatically stop and do not reject (accept) H_0. Let's look at the two examples from above:

Example 1
$\overline{X}_{RevSh} = 88$
$\overline{X}_{No\ RevSh} = 85$
$z = +2.00$

Example 2
$\overline{X}_{RevSh} = 72$
$\overline{X}_{No\ RevSh} = 81$
$z = -6.00$

In Example 1, the review-sheet mean of 88 is larger than the no-review-sheet mean of 85, so the means are in the direction predicted by the researcher. We can continue on to step two and conduct the z-test. In Example 2, the review-sheet mean of 72 is not larger than the no-review-sheet mean of 81, so we stop here and do not reject (accept) the null hypothesis.

Once you have decided that the means are in the direction predicted by the researcher, as they were in Example 1 (but not in Example 2), you can continue with the next step, which involves conducting the z-test. In this case (as with a two-tail test) the sign of both the calculated and critical z-values is irrelevant. All that you need in order to reject H_0 is for the value you calculate to be larger than the critical value of 1.65 (at the .05 level) or 2.33 (at the .01 level). So, in Example 1, the value of 2.00 is larger than the critical value of 1.65 so we can definitely reject H_0 at the $p < .05$ level. We cannot do any better than this (i.e. reject H_0 at the $p < .01$ level) because 2.00 is not larger than 2.33 (the critical value at the .01 level).

Remember that the first method for conducting a one-tail test involves subtracting the two sample means in the direction predicted by the alternative (research) hypothesis and comparing this calculated value to either the critical value in the positive or negative tail, depending on the research hypothesis. The sign of the calculated and critical z-values *does* matter when you conduct the z-test by this first method. The second method for conducting a one-tail test involves a two-step process in which you first verify that the means are in the direction predicted by the researcher and, if they are, you proceed to compare the calculated to the critical z-value regardless of the sign. The signs of the calculated and critical z-values do *not* matter when you conduct the z-test by this second method. These two methods will always yield exactly the same results, but you may prefer the second method as slightly easier because you do not have to worry about the sign (just as you don't worry about the sign with the two-tail test). But feel free to use whichever method you prefer (or both, in order to check yourself, if you are so motivated).

actual scores   two-tail   1.96 → 95%   2.58 → 99%

one-tail   1.65 → 95%   2.33 → 99%

Power and Error

So, why again do we want to do a one-tail test if a two-tail test allows us to test a difference between sample means in both directions whereas the one-tail test only allows us to compare a mean difference in the predicted direction? The answer we stated above was power. A one-tail test is more powerful than a two-tail test. How is this? **Power** refers to the ability to reject the null hypothesis. As we stated previously, a researcher hopes to reject the null hypothesis and accept the alternative (research) hypothesis, since the research hypothesis is what is being predicted. For example, as professors, we might hope to reject the idea that the review sheet does not help students perform better on the exam and to accept the idea that it does help them learn the material better. There are two major ways in which we can increase the ability to reject the null hypothesis. One of these ways is to conduct a one-tail test. Let's consider an example concerning the review sheets. Remember that a more powerful test is one that allows us to reject the null hypothesis over one that does not. We obtain the following sample means:

$$\overline{X}_{\text{No RevSh}} = 84.5$$
$$\overline{X}_{\text{No RevSh}} = 82$$

The raw difference score is $d = 84.5 - 82 = 2.5$ points. Looking at the distribution of differences we see that this difference falls 1.66 *SD* units above the mean, or at a $z = +1.66$:

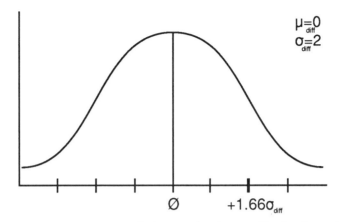

Figure 5.8 Distribution of differences with $\sigma_{diff} = 1.5$ and z-score difference of $z = +1.66$.

We could also have used the z-score formula to obtain this z-score value:

$$z = \frac{(\overline{X}_1 - \overline{X}_2)}{\sigma_{diff}} = \frac{84.5 - 82}{2} = \frac{2.5}{2} = 1.66$$

Now let's conduct the z-test considering this first as a two-tail problem (with a nondirectional research hypothesis) and then as a one-tail problem (with a directional research hypothesis). Let's start with conducting the z-test at the $p < .05$ level using a two-tail test:

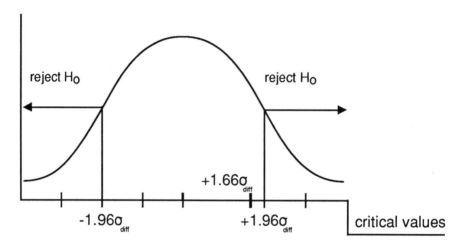

Figure 5.9 Rejecting the null hypothesis using a two-tail, non-directional hypothesis test at the $p < .05$ level

Since the calculated value of 1.66 is not larger than the critical value of 1.96, we do not reject (accept) the null hypothesis. There is no difference between those students who got the review sheet and those who did not. (Note that we do not bother to conduct the z-test at the $p < .01$ level if we cannot reject at the .05 level. That is, if you can't be at least 95% sure of your decision you certainly can't be 99% sure. The .01 level test will always have a higher critical value than the .05 level test so it will always be more difficult to reject H_0 at the .01 level. This is true for all statistical tests).

Now let's try this a second time but consider this as a one-tail problem instead. Again, let's start with conducting the z-test at the $p < .05$ level using a one-tail test:

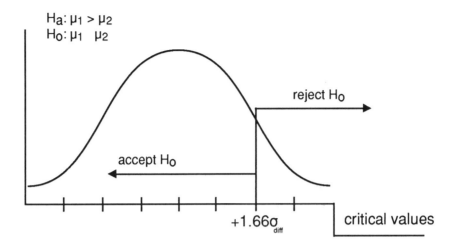

Figure 5.10 Rejecting the null hypothesis using a one-tail, directional hypothesis test at the $p < .05$

First, are the means in the direction the researcher predicted for the one-tail test? Yes -- the review-sheet mean of 84.5 is higher than the no-review-sheet mean of 82. Now let's compare the calculated and critical values: the calculated value of 1.66 is larger than the critical value of 1.65, so we can reject the null hypothesis at the $p < .05$ level. Wow! The one-tail test really is more powerful than the two-tail test. When we conducted the z-test as a two-tail test we could not reject the null hypothesis, but when we conducted the z-test at the one-tail level we did reject H_0. So, the one-tail test is more powerful than the two-tail test (but remember: with the one-tail test what you give up is the ability to test the significance of mean differences in the direction opposite than predicted in the alternative hypothesis). Note that in doing this one-tail test we chose the second method for conducting the test, but you could just as easily have used the first method by subtracting the means in the order: review-sheet minus no-review-sheet mean and then referred to the graph with the positive tail (see Figure 5.10).

So how do you know whether to conduct the more powerful one-tail test and look for mean differences in only one direction or conduct the less powerful two-tail test but check for mean differences in both directions? The answer to this question really depends upon the researcher and the question s/he is asking. A good rule of thumb is that if you are not very sure what is going to happen, but you think *something* will happen, that a two-tail test is probably more appropriate because you can test for the direction of influence. If you are fairly certain that if there is any effect it will be in one particular direction, and you really want the most ability possible to detect this difference, then a one-tail test is probably most appropriate.

So, a one-tail test is more powerful than a two-tail test. But we also mentioned there was a second way to increase power. This second way is to increase the number of subjects in each of the two samples. As it happens, this concept is not particularly relevant for the z-test (because you have perfect knowledge of the population to begin with). But, for the test that you will typically use to compare differences between two sample means, the t-test, this concept will become very important, so, we will hold off discussing this idea further until the next chapter. But, in the meanwhile, remember that the two ways to increase power (the ability to reject the null hypothesis) is to conduct a one-tail test over a two-tail test and to increase the sample size.

Before leaving this chapter we want to turn to one last concept: error. Know it or not we have been talking about error all throughout this chapter. When you reject a null hypothesis at the $p < .05$ level, we said that you are 95% sure that the observed difference in means is significant (true and meaningful and caused by the researcher's manipulation, such as presence of a review sheet). There is, however, a 5% chance that in fact the observed difference is purely due to chance, since 5% of all mean differences from the same population will fall outside the critical range of +/-1.96. This 5% is called error, or more specifically, **Type I error**. With a Type I error of .05 there is a 5% chance that you incorrectly rejected the null hypothesis, and concluded there was a significant difference between means when in fact the null hypothesis should have been accepted, and there was not a significant difference between means. One reason we try to reject H_0 at the $p < .01$ level if we can is because the Type I error is smaller. When H_0 is rejected at the .01 level, the Type I error is only .01, meaning that there is only a 1% chance that we incorrectly rejected the null hypothesis and a 99% chance that we made the correct decision. Type I error is especially problematic and researchers are most concerned about keeping this to a minimal level.

But what about the opposite case? What if you conclude that there really is not a significant difference between sample means when in fact there really is. For example, consider the two-tail case with a z-score of $z = 1.66$. With a critical value of 1.96, the statistical decision was to not reject (accept) H_0 because we could not be at least 95% confident (Type I error of only .05) that the review sheet did in fact cause those who used it to score differently than those who did not. But say that a value of 1.75 meant that we could be 90% confident that the difference in means was true and significant (Type I error of .10, a 10% chance of being wrong). In other words, there is a 90% chance that the review-sheet group truly was different from the no-review-sheet group, and only a 10% chance that these two sample means actually represent the same population. By not rejecting (accepting) the null hypothesis, if the alternative hypothesis actually were true, we would have committed an error, specifically, a Type II error. If the alternative hypothesis actually is true but you accept the null hypothesis, then we have committed a **Type II error**. Researchers generally consider Type II error slightly less of a problem than Type I error, because it is better to conclude that nothing is happening (when in fact something actually is) than to conclude that something really is happening (when in fact it is not).

The following table provides an illustration that you may find helpful in distinguishing the difference between Type I and Type II error:

		Researcher's Decision	
		Reject H_0	Accept H_0
In Reality	H_0 True	Type I	No Error
	H_0 False	No Error	Type II

Table 5.3 Decisions regarding Type I and Type II error

Note that when you reject the null hypothesis, when in fact the null hypothesis is false (and the research hypothesis is true), then you have not made a mistake. Similarly, if you do not reject the null hypothesis and in fact the null hypothesis is true (and the research hypothesis is false), then you have not made a mistake. Most of the time you hope not to make any mistakes, and you can keep any mistakes at a minimum by always requiring at least a .05 significance level to reject the null hypothesis.

Chapter 5 Problems

1. Define the null hypothesis (H_0) and the alternative (research) hypothesis (H_a).

2. Which hypothesis (*null* or *alternative*) is the researcher's prediction about the relationship between variables?

3. Which hypothesis (*null* or *alternative*) is tested statistically?

4. For each of the following symbolic representations, state which hypothesis (*null* or *alternative*) is represented below:
   a. $\mu_1 = \mu_2$
   b. $\mu_1 \neq \mu_2$

5. For each of the following pairs of hypotheses, identify which is the null hypothesis (H_0) and which is the alternative hypothesis (H_a). Also indicate whether this is a one- or two-tail hypothesis.
   a. $\mu_1 \geq \mu_2$ and $\mu_1 < \mu_2$
   b. $\mu_1 \neq \mu_2$ and $\mu_1 = \mu_2$
   c. $\mu_1 > \mu_2$ and $\mu_1 \leq \mu_2$

6. For each of the following, indicate whether the statement represents a null hypothesis or an alternative (research) hypothesis:
   a. Women will be more likely to show support to a distressed stranger than will men.
   b. There will be no difference between younger and older children's hand-eye coordination.

c. College seniors will express a different opinion on the issue of a tuition increase than will college freshman.
d. There will be fewer sensational stories published in the *New York Times* than in the *Washington Post*.
e. There will be no difference among employees' and managers' knowledge of company goals.

7. For each of the following, formulate an appropriate null hypothesis and alternative (research) hypothesis. State each hypothesis both verbally and symbolically:
   a. A researcher wants to examine whether there is a difference between university students' and professors' preference for a semester system.
   b. A researcher wants to determine whether families living in poverty have more children than do families living above the poverty bracket.
   c. A researcher wants to examine whether there is a difference in the number of crimes committed by men versus women.
   d. A researcher wishes to determine if there is a difference in the number of times students contact home for money if they are enrolled at private versus public universities.

8. For each of the following pairs of weather forecasts, which weather condition is the *least* likely to occur (the probability values represent how likely it is that the weather condition will occur)?
   a. rain, $p = .08$ or   snow, $p = .01$
   b. windy, $p = .45$ or   overcast, $p = .25$
   c. foggy, $p = .50$ or   sunshine, $p = .80$

9. For each of the following sets of three horses, which horse is *most* likely to win its race (the probability value represents the likelihood of that horse winning)?
   a. Lucky Shot, $p = .44$; Royal Blue, $p = .12$; French Admiral, $p = .51$
   b. Nobility, $p = .25$; Sapphire, $p = .70$; Gone with the Wind, $p = .68$
   c. Majestic Prince, $p = .99$; Good Fortune, $p = .81$; Victory, $p = .40$

10. What are the two most common probability levels? Describe what they mean?

11. Which test (*one-tail* or *two-tail*) should be conducted if the researcher *cannot* predict the direction of the mean differences between groups?

12. Which test (*one-tail* or *two-tail*) should be conducted if the researcher *can* predict the direction of the mean differences between groups?

13. For each of the following hypotheses indicate whether a one-tail or a two-tail z-test should be conducted:
   a. Students with GPAs over 3.5 will study more hours than will students with GPAs below 3.5.
   b. There will be a difference in the number of freeway deaths on highways where the speed limit is 55 mph compared to freeways where the speed limit is 65 mph.
   c. Voters will be more likely to vote for male political candidates than for female political candidates.
   d. Younger and older adults will differ in the amount of coffee they drink per day.

14. What is the critical value for deciding whether to reject the null hypothesis for a normal distribution when you conduct a:

a. two-tail test at the $p < .05$ level?
b. two-tail test at the $p < .01$ level?
c. one-tail test at the $p < .05$ level?
d. one-tail test at the $p < .01$ level?

15. When the calculated value is larger than the critical value the researcher should (*reject* or *not reject*) the null hypothesis.

16. For each of the following calculated z-scores (σ_{diff}), indicate whether the null hypothesis should be rejected or not rejected (accepted), assuming a $p < .05$ rejection level and a two-tail hypothesis:
    a. +1.65
    b. -2.01
    c. +1.50
    d. +2.68
    e. +1.33
    f. -0.75
    g. -3.00

17. For each of the following z-scores (σ_{diff}), indicate whether the null hypothesis should be rejected or not rejected (accepted), assuming a $p < .05$ rejection level and a one-tail hypothesis such that $\overline{X}_1 > \overline{X}_2$ (the difference between the means is positive):
    a. +1.85
    b. -0.50
    c. -4.00
    d. +2.00
    e. +0.75
    f. -1.90
    g. +1.75

18. Which error is committed when the researcher rejects the null hypothesis when in fact it should have not been rejected (i.e. when it is really true)?

19. If a researcher concludes that two samples are really from the same population (not significantly different from one another) when in fact the two samples really are from different populations, which error has the researcher committed?

20. What is the Type I error if the null hypothesis is rejected at the $p < .05$ level? at the $p < .01$ level?

21. Greater power means that the researcher has (*more* or *less*) ability to reject the null hypothesis?

22. Consider the following z-distribution for differences in IQ scores for a population of highly gifted individuals. The standard error of the difference is 5 points. A researcher shows up with two samples: the first sample has a mean IQ score of $\overline{X}_1 = 135$ and the second sample has a mean IQ score of $\overline{X}_2 = 115$. If you had to decide, with at least a 95% chance of being correct, whether or not both samples came from the same population of highly gifted people, what would your conclusion be? (Hint: conduct a two-tail z-test for this problem).

CHAPTER 6
THE T-TEST

From z To t Is Easier Than ABC

inferential statistics

Relative to the last chapter, this one should be a breeze. In fact, in Chapter 5 you already learned virtually everything you need to know in order to conduct your second statistical test, the t-test. The t-test really is nothing more than a z-test with a modification for different sample sizes. The purpose of the t-test is exactly the same as the purpose of the z-test: to ascertain whether two groups are significantly (not due to chance) different from one another. As with the z-test, the t-test requires interval- or ratio-level data in order to compute means for the two groups (samples). And, the formula for computing the calculated value from the two sample means is exactly the same as what you learned in the last chapter:

t-test - used w/ smaller sample size

$$t = \frac{(\overline{X}_1 - \overline{X}_2) - 0}{\sigma_{diff}} \text{ or } \frac{(\overline{X}_1 - \overline{X}_2)}{\sigma_{diff}}$$

Notice that this is the formula for computing a z-score for the distribution of differences-it is exactly the same formula for the t-test. Because this is the t-test, the computed values are called t-scores rather than z-scores. So how does the t-test differ from the z-test?

Well, you may have wondered in the last two chapters how anyone could ever take an infinite number of random samples from a population in order to create the distribution of differences or the sampling distribution. If you wondered this, so have we. In fact, if you already have perfect knowledge of a population (that is, you know how all 1,500 students in the population of introductory statistics students scored), then you will be able to create a normal distribution of differences (by exhaustively drawing random samples and plotting difference scores) with the critical values from the last chapter (i.e., +/- 1.96, +/-2.58, + or -1.65, + or -2.33). However, most of the time you do not have perfect information about everyone in the population. Consider a population of all American voters or all college students or all women. It would be difficult or impossible to find out the opinion or score on some dependent measure for every single person in these populations. So, if you can't use everyone in the population to create a distribution of differences (the z-distribution), how do you do it?

    The answer is that you use the two samples you are comparing to approximate the z-distribution: these approximations are called t-distributions. The larger the size of the two samples, the more of the population that is represented by the samples, so the closer the t-distribution comes to being a z-distribution. Let's see how this works using an example from a homework problem from a few chapters ago. Consider the population of all NFL football players (let's assume there are 1,000 players per team on 28 teams; not true, but we're taking a little creative license as the authors). So, the population we are dealing with consists of 28,000 football players. Suppose we are interested in the weight (the dependent variable) of NFL players, which is measured on a ratio-level scale. From the earlier chapter on distributions, we know that we could create the distribution of differences by drawing pairs of random samples of some size (10, 50, 100 or whatever) from the entire population, subtracting the means, and plotting the difference scores. But to do this we would need to find out the weight of every single one of the 28,000 players (since they would each eventually be sampled) -- an ominous task at best.

    Instead, let's take two samples of some specific size (N) and use these samples to approximate a distribution of differences. We already know that all distributions of differences will have a mean equal to zero, so all we need to worry about calculating is a standard deviation for this distribution. Since each of the two samples of size N have a standard deviation, we can use these two pieces of information to compute the standard error of the difference for the distribution of differences. Now, in essence, we have created a distribution of differences with a mean (equal to zero) and a standard deviation (the standard error of the difference, which is a function of the two sample standard deviations and the sample size), and we did this without drawing exhaustive random samples from everyone in the population (all 20,000 NFL players).

So, with a distribution of differences can't we just conduct a z-test to compare the sample means? The answer is no and the reason is that the distribution of differences we created may not be normally distributed. If you recall, one of the distinguishing features of a normal distribution is that the range has a constant relationship with the standard deviation: approximately 6 *SD* units comprise the range (difference between the high and low scores) of the distribution. That is, almost all of the distribution (99%) fits within +/-3 (actually +/-2.58) standard deviations around the mean. However, depending upon the size of the two samples that we used to approximate the distribution of differences, we may not have created a normal distribution that maintains this constant relationship. Rather, we created a distribution that is slightly more spread out (flatter, like a pancake) than a perfect normal distribution:

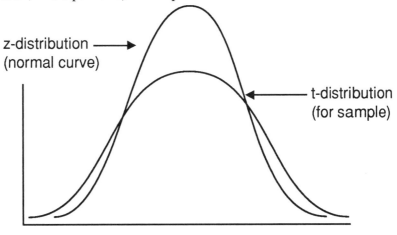

Figure 6.1 Comparison of z-distribution with t-distribution

How do we know how spread out this distribution of differences will be? That is, how different from a normal curve will the distribution be? Well, the answer to this is that it depends on the sample size. For each sample size, there is a unique t-distribution that corresponds with it. The smaller the sample size, the flatter the t-distribution; the larger the sample size, the more the curve (t-distribution) approximates a normal distribution. Generally, once you are working with samples of approximately size 1,000 per sample or larger, the t-distribution so closely approximates the z-distribution (because you have selected a pretty large sample out of the population) that you essentially have a z-distribution and are conducting the z-test. Let's look at some t-distributions for different sample sizes:

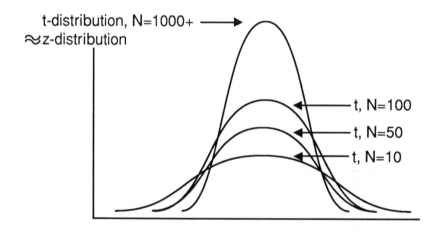

Figure 6.2 An example of t-distribution with increasing sample size

As you can see, the distribution of differences that was created using two samples each of size $N = 10$ is pretty flat and spread out. The t-distribution created using samples of size $N = 50$ and $N = 100$ are each slightly closer to the normal distribution. Finally, the t-distribution with samples of size $N = 1,000$

is superimposed over a normal z-distribution because they are so close to being almost the same.

So why is the t-distribution so spread out for small sample sizes? Well, let's consider an example with two different sample sizes: $N = 2$ and $N = 10$. We know from above that the t-distribution created with a sample size of $N = 2$ should be more flat (less like a normal curve) than should the t-distribution with a sample size of $N = 10$. That is, the range of the $N = 2$ t-distribution should be larger (to make the curve more spread out) than should the range of the $N = 10$ t-distribution. A large range on a distribution of differences would mean that there is a very large difference between two sample means. Now that we know this is how it is supposed to work, let's see why it actually does work this way (why the difference between two sample means will be larger with small sample sizes than will be the difference with larger sample sizes).

Consider a NFL player, Joe, who plays for the Buffalo Bills and weighs 425 pounds. Joe happens to be in one of the two samples that we used to create the distribution of differences with sample size $N = 2$:

Sample One		Sample Two	
Joe	425 lbs.	Brad	235
Rich	215 lbs.	Art	200
$\overline{X}_1 =$	320 lbs.	$\overline{X}_2 =$	217.5

The difference between these two sample means is $d = \overline{X} - \overline{X} = 320 - 217.5 = 102.5$. The standard error of the difference (the standard deviation for this distribution) will be quite large because, due to Joe, these weights are spread out pretty far, on average. You can see this reflected somewhat in the large difference score between the two sample means. Let's see what happens when Joe gets averaged in with a larger sample size $N = 10$:

134 THE T-TEST

Sample One		Sample Two	
Joe	425 lbs.	Brad	235 lbs.
Rich	215	Art	200
Alec	175	B.B.	265
Chris	220	Blake	160
Andy	235	Randy	225
Mike	210	Kalid	230
Ben	240	James	215
Walt	210	Achim	215
Wil	190	Calvin	220
Herb	200	Bud	210
$\overline{X}_1 =$	232	$\overline{X}_2 =$	217.5

Now, the difference between these two sample means is much smaller: $d = 232 - 217.5 = 14.5$ And, when a standard deviation (σ_{diff}) is computed, this figure will be quite a bit smaller than the standard error of the difference with the smaller sample size. In general, the larger the sample size the smaller the standard deviation for the distribution of differences, and the smaller the sample size the larger the standard deviation for the distribution of differences. What happens with larger sample sizes is that extreme scores (such as Joe, who weighs a couple hundred pounds more than everyone else) lose some of their power to influence the statistics when they get averaged in with a lot of scores (in large samples) than when they are averaged in with very few scores (in small samples).

Critical Values and The t-Distributions

As you may have surmised, if you are working with small sample sizes (smaller than about 1,000) you cannot refer to the critical values for the z-distribution. Remember that the critical values are the values you need to exclude the 5% or 1% outermost extreme scores in a distribution. The value of the calculated test statistic (computed from the two sample means) must exceed this critical value in order to reject the null hypothesis and conclude that the two groups are significantly different from each other. Let's review what this looks like for the two-tail z-test at the $p < .05$ level and the $p < .01$ level:

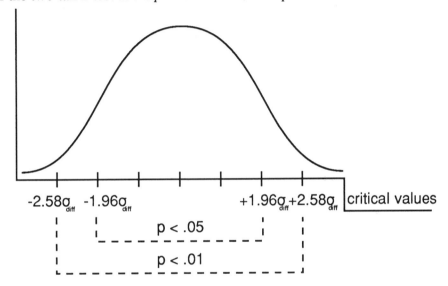

Figure 6.3 Rejecting the null hypothesis using a two-tail, non-directional hypothesis test at the $p < .05$ and $p < .01$ levels

Recall that the critical value for rejecting H_0 at the $p < .05$ level (excluding only the outermost extreme 5% of all difference scores) was +/-1.96, and for rejecting H_0 at the $p < .01$ level (excluding the outermost extreme 1% of all difference scores) the critical value was +/-2.58. Let's see what would happen with a t-distribution of size $N = 2$:

136 THE T-TEST

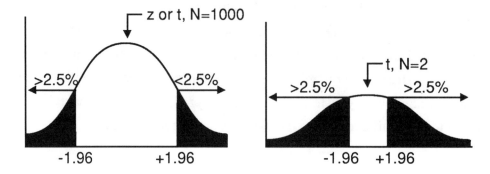

Figure 6.4 The percentage (shaded) of a z-distribution vs. a t-distribution of size $N = 2$ excluded by the z-values +/-1.96

As you can see, with a critical value of +/-1.96 we will end up excluding more than the 5% most extreme scores on the t-distribution. Obviously, then, these would not be the critical values for the $p < .05$ level. But we still want to exclude the extreme outermost 5% of scores on this t-distribution. To do this, we move the critical values out further (away from the mean) until we reach the point at which we are excluding the extreme outermost 5% (2.5% in each tail) of the distribution:

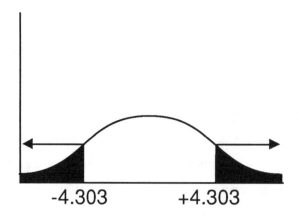

Figure 6.5 Critical value needed to exclude the extreme outermost 5% of a t-distribution with sample size $N = 2$

As it happens, with a sample size of $N = 2$, the critical value occurs at +/- 4.303. So, if the difference you calculate between two means, each with sample size $N = 2$, is larger than +/-4.303 (with a two-tail hypothesis) then you would reject the null hypothesis at the $p < .05$ level and conclude that the two means are significantly different from one another.

You should be able to guess by now that as the size of the sample increases (and as the distribution gets closer and closer to the normal z-distribution) the critical values needed to reject the null hypothesis will get smaller (closer to +/-1.96). Again, let's see how this would work with two sample means each of size $N = 10$:

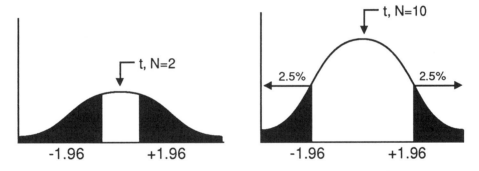

Figure 6.6 The percentage (shaded) of a z-distribution vs. a t-distribution of size $N = 10$ excluded by the z-values +/-1.96

Because the size of each sample is larger ($N = 10$ vs. $N = 2$ in the t-distribution will be closer to a normal z-distribution. Basically, this t-distribution with sample size $N = 10$ is less spread out than is the t-distribution with $N = 2$, so the critical values needed to exclude the extreme outermost 5% of the distribution are smaller with the larger sample size: (see Figure 6.6). With $N = 10$, the critical value needed to reject H_0 at the $p < .05$ level (with a two-tail test) is +/- 2.101. As you can see, this critical value is much closer to the critical z-value of

+/-1.96 than is the critical value for the t-distribution with sample size $N = 2$ (+/-4.303).

You may recall from the last chapter that we said there were two ways to increase the power of the z-test. Well, technically, there are two ways to increase the power of the t-test. One you already know, which is to move from a two-tail to a one-tail test because you put the entire 5% (or 1%) error into only one tail of the distribution rather than dividing it up into both sides. Recall, however, that what you give up with the one-tail test is the ability to test for a difference in means in the opposite direction than predicted. The second way to increase power we have just illustrated above. Remember that power refers to the ability to reject the null hypothesis, and the null hypothesis is rejected whenever the calculated value (computed using two sample means) is larger than the critical value (which excludes the extreme outermost 5% of the t-distribution for $p < .05$ rejection or 1% for $p < .01$ rejection). With a sample size of $N = 2$, we would need a calculated value larger than +/-4.303 in order to reject H_0, whereas with the larger sample size of $N = 10$ we would need a calculated value larger than only +/-2.101. So, we have increased the ability to reject H_0 with a larger sample size: we have more power! This exact same principle applies if you are looking at the $p < .01$ level or if you are looking at a one-tail test: the critical value will be smaller (closer to the critical z-value) if you have a larger sample size.

Degrees of Freedom and the t-table

So, how do you know what the critical values are for the various t-distributions? How did we know that for a sample size of $N = 2$ the critical value was +/-4.303 for a two-tail test at the $p < .05$ level, and so forth? Well, the answer is that for each sample size there is a unique t-distribution that has specific critical values that have been compiled into a table known as the **t-table** (see Appendix A). This table shows the critical values at the $p < .05$ and $p < .01$ levels for the one-tail and two-tail t-tests. There are a couple important features to note about the t-table. One is that the t-table does not list the critical values by sample size as we were discussing above, but rather it lists critical values according to degrees of freedom. Degrees of freedom is directly related to sample size. **Degrees of freedom** refers to how many scores are free to vary in the statistical test in order to obtain the observed values. Here's how this works. Consider a sample with five people (five scores) and a mean of 80. Basically, we are free to vary four of these scores to be whatever we want, but once we have selected these four scores the fifth score is automatically predetermined in order to obtain the specified mean of 80. For example, let's suppose we pick miscellaneous scores for the first four people as follows:

Example One

Ben	81
Sue	73
Johann	90
Marcy	88
Merle	??
$\overline{X}_1 =$	80

In order to obtain a mean of 80 there is only one possible value that Merle's score can be, and that value is 68. If Merle were to have any other score the mean for all five students would not equal 80.

Again, note in this second example that we can vary four of the scores to be anything, but once these are selected the final score is predetermined:

Example Two

Ben	79
Sue	86

Johann	??
Marcy	92
Merle	69
$\overline{X}_2 =$	80

Notice that Johann's score is predetermined to be only one possible score -- 74. There are four degrees of freedom ($df = 4$) with a sample size of $N = 5$. So the general rule is that you are free to vary all but the last score. So, with a sample size of $N = 20$, there would be $df = N - 1 = 20 - 1 = 19$ degrees of freedom.

Because with a t-test we are always comparing two samples, we lose one degree of freedom for each sample. Consider a t-test with two samples of size $N_1 = 2$ and $N_2 = 2$. We lose one degree of freedom for each sample, so $df = (N_1 - 1) + (N_2 - 1) = (2 - 1) + (2 - 1) = 1 + 1 = 2$ degrees of freedom. If we use this value of df = 2 (corresponding to two samples each of size $N = 2$) and look on the t-table we find the critical value for a two-tail t-test conducted at the $p < .05$ level of +/-4.303. What are the degrees of freedom with two samples each of size $N = 10$? Again, we lose one degree of freedom for each one of the two samples, so $df = (N_1 - 1) + (N_2 - 1) = (10 - 1) + (10 - 1) = 9 + 9 = 18$ degrees of freedom. A two-tail t-test with 18 degrees of freedom at the $p < .05$ level requires a calculated value larger than the critical value of +/-2.101. What would the critical value be if this test were conducted at the $p < .01$ level and was a one-tail test? The answer is 2.552 (either positive or negative, depending on which tail is being used). Take some time to familiarize yourself with the t-table so you feel comfortable finding critical values for various t-distributions with different degrees of freedom.

A second important thing to notice about the t-table is the critical values that correspond to an infinite (∞) number of degrees of freedom (essentially this is reached once you have samples of size 1,000 or larger). Do these numbers look at all familiar? They should -- they are the critical values for the z-test! Recall that we said that t-distributions are approximations for the theoretical z-distribution for different sample sizes. Once you have samples of approximately size $N = 1,000$ or larger, then the t-distributions are so close to normal that they virtually are the z-distribution. You already know these critical values: a two-tail test at the $p < .05$ level has a critical value of +/-1.96; a two-tail test at the $p < .01$ level has a critical value of +/-2.58; a one-tail test at the $p < .05$ level has a critical value of +1.65 or -1.65; and a one-tail test at the $p < .01$ level has a critical value of +2.33 or -2.33. So the z-test and t-test really are closely related!

An Example of the t-Test

Let's try conducting an example of the t-test. Suppose a researcher (or your statistics professor) wants to test the effectiveness of a new teaching guide that is designed to help students understand the course material better. The researcher decides that to conduct this test she is going to use the statistics class you are enrolled in. The population of interest is all students taking an introductory statistics class. Because your class is convenient, she decides to use it to conduct her study (an experiment). She randomly assigns half of the class ($N = 12$) to use the study guide (the experimental group) and randomly assigns the other half of the class ($N = 12$) to not use the study guide (control group). At the end of the quarter, the researcher calculates the mean score on the final exam for those in the experimental group and those in the control group. The experimental group had a mean $\overline{X}_E = 90$ on the final exam and the control group had a mean $\overline{X}_C = 82$ on the final exam. The standard error of the difference (the standard deviation for the distribution of differences), which is calculated based on the size of the two samples (12 each) and the standard deviations for each of the two samples, turns out to be $\sigma_{diff} = 3$.

QUESTION: Did the study guide cause students to score significantly (not due to chance) higher than students who did not use the study guide?

ANSWER:

Step 1: Identify the hypotheses for this study.

Since we know we need to test the null hypothesis, we should first determine what it is. The researcher is predicting that the study guide will *improve* students' understanding of the course material, so the research hypothesis is:

H_a: Students who use the study guide will score higher on the final exam than will students who do not use the study guide.

The corresponding null hypothesis (the opposite of the research hypothesis) is:

H_0: Students who use the study guide will score no differently (or lower) on the final exam than will students who do not use the study guide.

Stated symbolically, these hypotheses look like the following:

H_a: $\mu_E > \mu_C$
H_0: $\mu_E \leq \mu_C$

Since the researcher is predicting that the means will fall in a particular direction, we know that we will be conducting a directional or one-tail t-test.

Step 2: Calculate the value for t.

Now the researcher needs to determine what type of statistical test she should be conducting. First, since we are working with means (obtainable only from interval- or ratio-level data) and because we are comparing two groups, we must conduct either a z-test or a t-test, but which one? Well, because the researcher does not have access to all possible statistics students in the world (which she would need in order to create a distribution of differences for a z-test), she knows she must approximate the z-distribution based on the two sample means. So, we will be conducting a t-test.

Remember that the t-score formula is exactly the same as the z-score formula for the distribution of differences:

$$t = \frac{(\overline{X}_1 - \overline{X}_2)}{\sigma_{diff}}$$

So, the t we would calculate using the two sample means and the standard error of the difference is:

$$t = \frac{(\overline{X}_1 - \overline{X}_2)}{\sigma_{diff}} = \frac{90 - 82}{3} = \frac{8}{3} = +2.66$$

The calculated value is $t = +2.66$. Is this difference significant; can we reject H_0 and conclude that those who used the study guide scored significantly higher than those who did not use the study guide?

Step 3: Calculate the degrees of freedom.

Because we need to look up a critical value to compare the calculated value to, we need some degrees of freedom value. Recall that one degree of freedom is lost for each of the two samples, so the degrees of freedom for this problem are:

$$df = (N_1 - 1) + (N_2 - 1) = (12 - 1) + (12 - 1) = 11 + 11 = 22$$

Step 4: Look up the critical value at the $p < .05$ level.

Since this is a directional, one-tail hypothesis we will be referring to the critical values on the t-table that correspond to a one-tail test. The critical value at the $p < .05$ level with 22 degrees of freedom for a one-tail t-test is: $t_{(22).05} = 1.717$.

Step 5: Compare the calculated t to the critical t.

Remember that when conducting the one-tail test to be cautious about the direction of the means (or the sign of the calculated t). With a two-tail test the sign does not matter since the t-test is nondirectional, but with a one-tail test the sign of the calculated value is important. Recall from Chapter 5 that we had two methods for conducting the one-tail test. Let's do both. First, let's try the method we prefer where the sign of the t is not relevant. This method involves first looking at the observed means and seeing if they are in the direction predicted by the researcher. If they are not, you can stop here and do not reject H_0. In this study, the researcher predicted that the study-guide group would have a higher mean than would the no-study-guide group. Is this what happened with observed means? Yes -- the study-guide group had a mean of $\overline{X}_E = 90$, which is indeed larger (in the predicted direction) than the mean of the no-study-guide group mean of $\overline{X}_C = 82$. Now we can proceed to compare the calculated and critical values without worrying about the sign of the calculated value and whether we are looking at the positive or negative tail of the t-distribution. Is the calculated value of $t = 2.66$ larger than the critical value of $t = 1.717$? Yes -- we definitely can reject H_0 at the $p < .05$ level, leaving a Type I error (chance of being wrong) of 05.

Let's do this same t-test using the second method for conducting the one-tail test. With this second method the sign of the t does matter. We need to be sure that when we computed the t we subtracted the means in the same direction as predicted in the research hypothesis. Since the research hypothesis predicted the study-guide-group mean would be larger than the no-study-guide group mean, we need to be sure to subtract the means in this same order: study-guide group mean minus no-study-guide group mean. As you can see from Step 2, we did subtract the means in the correct order and obtained a positive t value of $t = +2.66$. Since the researcher is predicting the first mean (study-guide group) will be larger than the second mean (no-study-guide group) she is predicting a positive difference, which means that we will be looking at the positive tail. The critical value is $t = +1.717$:

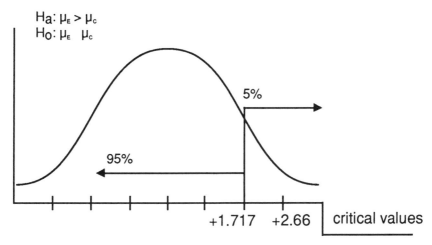

Figure 6.7 Rejecting the null hypothesis using a one-tail, directional hypothesis test

Since, the calculated value of $t = +2.66$ is larger than the critical value of $t = +1.717$, we can reject H_0 at $p < .05$. As you can see, the conclusion is always the same with the two methods for conducting the t-test and you can choose to use whichever method you prefer. So, now that we have rejected H_0 at $p < .05$ and concluded there is a significant difference (with a 5% chance of being wrong about this decision) do we stop here? No, we proceed to the $p < .01$ level to see if we can get the Type I error lower.

Step 6: Look up the critical value at the $p < .01$ level.

Looking at the one-tail section of the t-table under the $p < .01$ rejection level, we find a corresponding critical value for 22 degrees of freedom of $t_{(22).01} = 2.508$.

Step 7: Compare the calculated t to the critical t.

Since we already know we can reject H_0 at $p < .05$ we know the means are going in the predicted direction, so all we need to do now is see if the calculated t exceeds the critical t at the more stringent $p < .01$ level. Since the calculated value $t = 2.66$ is larger than the critical value $t = 2.508$, it falls in the extreme outermost 1% of the distribution:

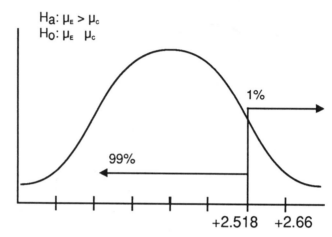

Figure 6.8 Rejecting the null hypothesis using a one-tail, directional hypothesis test

So, we can reject the null hypothesis at the $p < .01$ level. We can conclude that the experimental group scored significantly higher than the control group with a Type I error, or chance of being wrong, of only .01.

Step 8: State the final conclusion both statistically and in words.

Statistically, the final decision for this problem is to reject H_0 (the null hypothesis) at the $p < .01$ level. In words, the researcher concludes that the experimental group scored significantly higher than the control group with a Type I error (or chance of being wrong) of .01. The researcher is 99% confident that the study guide improves learning of course material. There is only a 1% chance that the study guide either does not help students at all or actually hurts their performance on exams. As a result of her findings for this study, the researcher decides to use her study guide in all of her future statistics classes.

We can take these basic steps we outlined above for conducting a t-test and put them into a flow chart as follows. You may find it useful to use this flow chart to help you work through the steps you need to follow to conduct the t-test and to know when to stop:

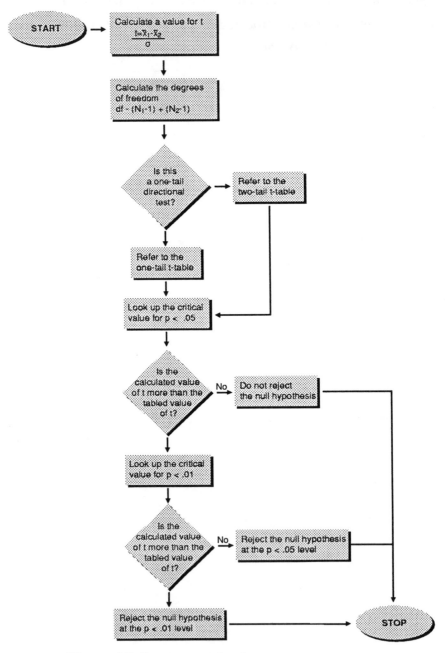

Figure 6.9 Steps to conducting a t-test

CHAPTER 6 PROBLEMS

1. What is the formula used for conducting the t-test?

2. What is the formula for finding the degrees of freedom for conducting a t test?

3. Which distribution (*sample, population, sampling,* or *differences*) is used to conduct a t-test?

4. What three pieces of information do you need in order to look up the correct critical value in the t-table?

5. When there is enough of a probability (.95 or higher) that two groups do not represent the same population, the researcher (*rejects* or *accepts*) the null hypothesis.

6. When there is enough probability (.95 or higher) that two groups do not represent the same population, the conclusion is that the two groups are (*significantly* or *not significantly*) different from one another.

7. For each of the following calculated t-values and sample sizes, indicate the degrees of freedom and whether you should reject or accept the null hypothesis (and at what significance level, if you reject). Use a two-tail hypothesis test.
   a. $t = +2.18$ $\quad N_1 = 5\ N_2 = 5$
   b. $t = -2.05$ $\quad N_1 = 12,\quad N_2 = 10$
   c. $t = +2.18,$ $\quad N_1 = 15,\quad N_2 = 15$
   d. $t = -2.05$ $\quad N_1 = 16,\quad N_2 = 16$

150   THE T-TEST

8. Look at your four answers in question #7. What can you conclude is the relationship between sample size and rejecting the null hypothesis (i.e., is the t-test more powerful with smaller or with larger sample sizes)?

9. How many subjects are there in each sample of these studies if the following degrees of freedom are from t-tests with *equal* sample sizes?

   a. 46 df
   b. 120 df
   c. 28 df
   d. 14 df
   e. 22 df
   f. 40 df

10. As a political consultant you hypothesize that people will react differently to a negative political ad than to a neutral ad. You have a total of 20 voters watch one of two commercials produced by the Democratic party. Ten people are randomly assigned to watch a commercial slandering Republicans and 10 people are randomly assigned to see a more neutral commercial. After watching the appropriate commercial, voters indicate their degree of liking for the commercial on a scale ranging from 1-7 with one as the lowest liking score and seven as the highest score. You obtain the following data:

   Negative Commercial
   $N_1 = 10$
   $\overline{X}_1 = 4$

   Neutral Commercial
   $N_2 = 10$
   $\overline{X}_2 = 6.5$

   The standard error of the difference between means is $\sigma_{diff} = 1.20$

   a. Is this a one-tail or a two-tail hypothesis?

   b. Formulate appropriate null and alternative hypotheses.

   c. Find $t_{calculated}$.

d. Find $t_{critical}$ at $p < .01$ and $p < .05$.

e. State your decision both statistically and in words.

f. If you were recommending which ad the Democratic party should show, which ad would you suggest (assuming that the more you like an ad, the more likely you are to vote according to the party sponsoring the ad)?

g. If you had hypothesized *instead* that voters who viewed a negative ad would like it less than would voters who viewed a neutral ad what would be the appropriate null and alternative hypotheses?

h. What would your statistical decision be for part g?

11. A sociology researcher is interested in whether a certain hour-long film which portrays the effects of racial prejudice will affect attitudes toward a minority group. Two groups of 31 subjects each were randomly assigned to one of two conditions. Group A watched the movie whereas Group B spent the hour playing cards. Both groups were then given a racial attitude test, with high scores representing a higher level of prejudice. The data are as follows:

Group A
$N_A = 31$
$\overline{X}_A = 42.60$

Group B
$N_B = 31$
$\overline{X}_B = 38.62$

The standard error of the difference between the means is $\sigma_{diff} = 1.36$.

a. Is this a one-tail or a two-tail hypothesis?

b. Formulate appropriate null and alternative hypotheses.

c. Conduct a t-test.

d. State your decision both statistically and in words.

12. A clinical sample of 122 delinquent boys was selected and randomly assigned to two groups. A clinical researcher was interested in discovering whether a six-week, nondirective, individual therapy program would affect levels of measured anxiety among delinquent boys. The boys in Group 1 all received the therapy, whereas those in Group 2 did not. Both groups were then given an anxiety-level test (high scores indicating more anxiety). The data were as follows:

Group 1
$N_1 = 61$
$\overline{X}_1 = 98.06$

Group 2
$N_2 = 61$
$\overline{X}_2 = 102.35$

The standard error of the difference between the means is $\sigma_{diff} = 2.45$.

a. Is this a one-tail or a two-tail hypothesis?

b. Formulate appropriate null and alternative hypotheses.

c. Conduct a t-test.

d. State your decision both statistically and in words.

13. A researcher hypothesizes that people who watch the *Brady Bunch* will want to have larger families than will people who watch *Three's Company*. The researcher randomly assigns 31 people to watch the *Brady Bunch* every night for one month and randomly assigns 31 other people to watch *Three's Company* every night for the same month. At the end of the month the participants are asked how many children they would like to have.

Brady Bunch	Three's Company
$N_{BB} = 31$	$N_{TC} = 31$
$\overline{X}_{BB} = 3.9$	$\overline{X}_{TC} = 1.8$

The standard error of the difference between means is $\sigma_{diff} = 0.82$

    a.   Is this a one-tail or a two-tail hypothesis?

    b.   Formulate appropriate null and research hypotheses.

    c.   Conduct a t-test.

    d.   State your decision both statistically and in words.

14. A researcher hypothesizes that men will initiate a conversation with a woman more quickly than with another man. The researcher randomly assigns 13 men to be placed alone in a room with a woman for one minute (the same woman is used for all 13 men). Another 13 men are randomly assigned to be alone in a room with a man for one minute (the same man is used for all 13 men). The researcher monitors the sound in the room and determines how long (in seconds) it takes male subjects to initiate a conversation with the other person in the room within the allotted minute. The following means are the length of time, in seconds, *before* the subjects initiated a conversation (i.e., a smaller number indicates that a conversation was initiated *more* quickly).

With Female In Room	With Male In Room
$N_F = 13$	$N_M = 13$
$\overline{X}_F = 18$	$\overline{X}_M = 12$

The standard error of the difference between means is $\sigma_{diff} = -1.45$.

    a.   Is this a one-tail or a two-tail hypothesis?

    b.   Formulate appropriate null and alternative hypotheses.

c. Can you conduct a t-test? Why or why not?

If the situation were as follows: $\overline{X}_1 = 5$, $\overline{X}_2 = 9$ and $\sigma_{diff} = 2.30$.

d. Conduct a t-test.

e. State your decision both statistically and in words.

CHAPTER 7
SINGLE-FACTOR ANALYSIS OF VARIANCE (ANOVA)

How ANOVA Differs From t

You can relax a bit now in knowing that the toughest part you covered was in Chapter 5 when you learned about hypothesis testing and significance. Fortunately, everything else from here on out is just a modification of what you have already done. This is particularly true for the next two chapters. You already know a good deal of what you need to conduct your next statistical test: the F-test or analysis of variance (ANOVA). An ANOVA or F-test is very closely related to the t-test, and we will show you how shortly, but take a moment to pat yourself on the back for getting through your first two statistical tests (z and t) and for mastering most of the basic concepts of hypothesis testing and conducting statistical tests! Congratulations!

Recall that a t-test (we will stop referring to the z-test from here on out because it is really a theoretical test that presumes perfect knowledge of the population, which you almost never have) is used when you have interval- or ratio-level data and when you are comparing two groups. One of the things we will emphasize in this and future chapters is that you need to be able to determine what statistical test you should be conducting. This will be particularly important to you if you ever do any type of research. The two primary things you need to concern yourself with are what level of measurement the dependent variable is (nominal, ordinal, or interval/ratio) and how many groups or conditions you have. The level at which data is measured will constrain what possible tests you can conduct (although advanced statisticians can manipulate data at one level and turn it into data at another level so they can conduct a particular test). The final determinant of what statistical test you will conduct is how many groups you have. Recall, again, that a t-test is conducted: 1) when you have interval/ratio level data, and 2) when you are comparing means for two, and only two, groups (e.g., a study-guide group and a no-study-guide group).

Let's extend this problem with the study guide from the last chapter. Suppose the researcher is also interested in testing whether extra help sessions she holds with her students helps them understand course material better and score higher on the exams. Now we actually have three groups (or at least we will view it this way for this chapter; we will see in the following chapter how we might view this problem instead as having four conditions). The three groups for this revised problem are: a study-guide group, an extra-help-session group, and a group that gets no study guide or extra help session (the control). Essentially, the experiment now has two experimental conditions and one control condition. The researcher decides to conduct the experiment to look at differences among these three groups in her statistics class the following semester. She randomly assigns one third of the class ($N = 15$) to use the study guide during the course, one third of the class ($N = 15$) to go to extra help sessions, and one third of the class ($N = 15$) to have no study guide or extra help. So, there are a total of 45 students who participated in this study. At the end of the semester, the researcher finds that the average final exam scores for students in each of the three groups was as follows: the study-guide group had a $\overline{X}_{SG} = 85$ on the midterm, the help-session group had a mean $\overline{X}_{HS} = 90$, and the control group had a mean $\overline{X}_C = 82$.

QUESTION: Is there a difference among these three groups in terms of how well they scored on the final exam?

How could we answer this question based on what we already know? Well, we do know how to compare two groups at a time to see if they are statistically different from one another? So, what we *could* do is conduct a t-test comparing the study-guide group to the control (we did this in the last chapter). Because we always start by conducting the test at the $p < .05$ level, we can assume a maximum Type I error associated with this test of .05 (even though in the last chapter we really ended up with a Type I error of .01 for the final problem, we didn't know that would be the final error since we started at the $p < .05$ level). So, we know so far we have a maximum 5% chance of making some wrong conclusion in answer to the question above. Now let's conduct a second t-test in which we compare the help-session group to the control. Again, we will be conducting the t-test with a maximum Type I error of .05. So, we now have conducted two t-tests, each with a possible Type I error of .05, so the total chance of making a wrong conclusion for this problem (in answer to the question above) is .05 + .05 = .10, or 10% chance of making a wrong decision (the Type I error has now compounded to .10). We have a third comparison to make: a t-test comparing the means for the study-guide group and the help-session group. Again, we will conduct this test with a maximum Type I error of .05. Added to the .10 error for the other two tests we end up with an error of .15 or a 15% chance of making a wrong decision about whether these three groups differ from one another. This is a fairly high error rate. We already noted that researchers really want to keep their chance of making a wrong statistical decision at the .05 level or lower. So how can we do this when we need to compare more than two group means?

The answer is a statistical test known as analysis of variance, or ANOVA for short. We actually will present two different forms of the ANOVA test (single- and multiple-factor) in this and the following chapter, but the basic idea is the same: you will be comparing means for more than two groups in a single statistical test. The rule for when to use analysis of variance: you have interval/ratio level data and you are comparing means for more than two groups (conditions). The single- vs. multiple-factor refers to the number of independent variables there are. A **factor** is exactly the same as an independent variable. If there is one independent variable (as there was with the t-test) then the problem is single factor. If there is more than one independent variable then the problem is multiple factor. For a t-test the single factor (one independent variable) can have only two levels or conditions. For a single-factor ANOVA or F-test the single factor (one independent variable) has three or more levels. For a multiple-factor ANOVA (which we will talk about in detail in the next chapter), there are two or more factors (at least two independent variables) with at least two levels each.

As with the t-test, the F-test has a research or alternative hypothesis, which is what the researcher predicts, and a null hypothesis, which is what is actually tested for statistical significance. When there were two groups, the researcher predicted that students in the study-guide group would score significantly different (in the nondirectional, two-tail scenario) than would students in the no-study-guide group. The corresponding alternative and null hypotheses were as follows:

$$H_a: \quad \mu_{SG} \neq \mu_C$$
$$H_0: \quad \mu_{SG} = \mu_C$$

Now, for the single-factor ANOVA consider the situation in which we have three groups. In this case, the researcher is predicting that not all three groups will score the same (that there will be some significant difference among the means of the three groups):

$$H_a: \quad \mu_{SG} \neq \mu_{HS} \neq \mu_C$$
$$H_0: \quad \mu_{SG} = \mu_{HS} = \mu_C$$

So, the hypotheses are very similar with the t-test and analysis of variance. One difference to note is that the concept of one- and two-tail does not come in to play with analysis of variance. Although the researcher may predict something directional, such as $H_a: \mu_{SG} > \mu_{HS} > \mu_C$ (that those who received either the study guide or the extra help session will score higher than those in the control condition), the corresponding null hypothesis for a directional research hypothesis is tested in exactly the same was as is a nondirectional hypothesis.

How Single Factor Analysis of Variance Works

Analysis of variance is a statistical test that compares means by analyzing variance in scores (hence, the name). That is, presumably not every single person will obtain exactly the same score on any given dependent measure (e.g., not everyone in a statistics class scores exactly the same on the final exam). So there is variability or variance in scores. When comparing groups to one another (as when we are trying to compare a study-guide group, a help-session group, and a control group), this source of variance among scores can come from two sources: differences between group means and differences in scores within a particular group. Let's examine each of these sources of variance more closely.

Consider the researcher's problem above: there are three conditions (study guide, help session, control) and each group or condition consists of a total of 15 scores (because there are 15 people per group). The data for final exam scores for each of the 45 students participating in this study might look something like the following:

Study Guide Group		*Help Session Group*		*Control Group*	
1. Barb	87	16. Tom	92	31. Mark	77
2. Warren	81	17. Burt	95	32. Chris	80
3. Andy	86	18. Cliff	88	33. Betsy	81
.....		.....		.....	
15. Corinne	82	30. Claire	89	45. Rudy	82

$\overline{X}_{SG} = 85$ $\overline{X}_{HS} = 90$ $\overline{X}_C = 80$

Grand Mean for all 45 students = $\overline{X}_G = 85$

As you can see, the scores vary in two ways. First, the scores vary between groups. You can see this by looking at how the three group means compare to the overall mean for all 45 students ($\overline{X}_G = 85$): $\overline{X}_{SG} = 85$, $\overline{X}_{HS} = 90$, and $\overline{X}_C = 80$. The means for the three groups were not always exactly the same as the grand mean. This is **variance between groups**. Second, the scores vary within groups. You can see this by looking at the scores within any particular group. In the study-guide group not everyone scored the mean of 85: Barb had an 87, Warren had an 81, and so forth. Similarly, looking at the help-session group, not everyone scored the mean of 90: Tom had a 92, Burt had a 95, and so on. This is **variance within groups**.

By analyzing these two sources of variance we can determine whether the three group means are significantly different from one another or whether the observed differences between groups are purely due to chance (how people were randomly assigned to groups; the study guide and help sessions had no influence on final exam scores). In particular, we can devise a statistic that measures these two sources of variance by looking at the differences in scores and group means. Let's do this for both variance between groups and variance within groups. First, for variance between groups there are three group means each consisting of 15 students scores: $\overline{X}_{SG} = 85$, $\overline{X}_{HS} = 90$, and $\overline{X}_C = 80$, and an overall grand mean for all 45 students (across the three groups), $\overline{X}_G = 85$. We can compute three differences as follows, comparing each group mean to the overall grand mean:

$$\overline{X}_{SG} - \overline{X}_G = 85 - 85 = 0$$
$$\overline{X}_{HS} - \overline{X}_G = 90 - 85 = +5$$
$$\overline{X}_C - \overline{X}_G = 80 - 85 = -5$$

Now, we could add up these three differences to get an overall measure of how far each of the group means deviates or varies from the overall grand mean: (0) + (+5) + (-5) = 0. However, this is not an accurate representation of the true variability of these scores; it looks as if there is no variability among these three group means at all (i.e., that they are exactly the same). As you can see this is not true. The problem is that we are adding positive and negative numbers which are canceling each other out and seemingly indicating no variability whatsoever. In order to get around this problem with the negative sign on some differences canceling out positive differences, we can simply take each of the differences and square them; this will eliminate negative numbers:

$$\overline{X}_{SG} - \overline{X}_G = 85 - 85 = 0 \qquad 0^2 = 0$$
$$\overline{X}_{HS} - \overline{X}_G = 90 - 85 = +5 \qquad +5^2 = 25$$
$$\overline{X}_C - \overline{X}_G = 80 - 85 = -5 \qquad -5^2 = 25$$

Now if we sum up these squared values we do obtain a figure which is a representation of the variability of the three group means: 0 + 25 + 25 = 50. Consider if the group means were even more variable: the study-guide group still had a mean $\overline{X}_{SG} = 85$, but the help-session group had a mean $\overline{X}_{HS} = 95$ and the control group had a mean $\overline{X}_C = 75$. The grand mean is still 85, but let's see what happens to our measure of variability:

$$\overline{X}_{SG} - \overline{X}_G = 85 - 85 = 0 \qquad 0^2 = 0$$
$$\overline{X}_{HS} - \overline{X}_G = 95 - 85 = -10 \qquad +10^2 = 100$$
$$\overline{X}_C - \overline{X}_G = 75 - 85 = +10 \qquad -10^2 = 100$$

Now if we sum the squared differences between each group mean and the grand mean we obtain the following: 0 + 100 + 100 = 200. This summed value of 200 for the second set of data is clearly larger than the summed value for the first set of data = 50. This makes sense because the group means are more variable in the second set. This sum of the squared differences between the group means and the grand mean (for looking at variance between groups) is not surprisingly called the **sum of squares between groups**. Statisticians generally try to come up with names for things that logically make sense (analysis of variance is named this because it is a test that analyzes the variance between and within groups; sum of squares is named this because it is the sum of squared differences).

As with between-groups variance, we also can compute a sum of squares for within-groups variance. In this case we are looking at the difference between each individual score and its own group mean. So, this computation involves summing up 45 differences (because there are 45 individual scores). Let's compute a sum of squares for each of the three groups, and then add them to form the total sum of squares within groups:

Study Guide Group		*difference*	*squared difference*
1. Barb	87	87 - 85 = +2	$+2^2 = 4$
2. Warren	81	81 - 85 = -4	$-4^2 = 16$
3. Andy	86	86 - 85 = +1	$+1^2 = 1$
.....		.....	.....
15. Corinne	82	82 - 85 = -3	$-3^2 = 9$
$\overline{X}_{SG} = 85$			Sum = 95

Help Session Group		*difference*	*squared difference*
16. Tom	92	92 - 90 = +2	$+2^2 = 4$
17. Burt	95	95 - 90 = +5	$+5^2 = 25$
18. Cliff	88	88 - 90 = -2	$-2^2 = 4$
.....		.....	.....
30. Claire	89	89 - 90 = -1	$-1^2 = 1$
$\overline{X}_{HS} = 90$			Sum = 130

Control Group		difference	squared difference
31. Mark	77	77 - 80 = -3	$-3^2 = 9$
32. Chris	80	80 - 80 = 0	$0^2 = 0$
33. Betsy	81	81 - 80 = +1	$+1^2 = 1$
.....		.....	.....
45. Rudy	82	82 - 80 = +2	$+2^2 = 4$
$\overline{X}_C = 80$			Sum = 75

Now, if we add up the three summed differences we obtained for each of the three groups: 95 + 130 + 75 = 300. This value represents the **sum of squares within groups** since it is the sum of the squared differences between each individual score and its own group mean.

We can take the sum of squares between value ($SS_B = 50$) and the sum of squares within value ($SS_W = 300$) and put them into a table to show the two sources of variance between and within groups:

Source	SS
Between	50
Within	300
Total	350

We also added a total line to tell the total variance or variability of scores represented by the sum of between groups variance and within groups variance; so the sum of squares total equals the sum of squares between groups and the sum of squares within groups: $SS_T = SS_B + SS_W$. The type of table we have started to create above is called an ANOVA table, and we will use this to work out the computation for the F-test (that is, to see if there are any significant differences among the three group means).

So, what should the sum of squares look like? In other words, what is the best case for detecting a significant difference? Well, ideally you want the differences between groups to be very large and the differences within groups to be very small. Let's consider a small example of three groups of five people each. In this example we are looking at the dependent measure of people's weight. We are interested in whether one group weighs significantly more or less than any other group. Consider the following two different scenarios for these three groups:

Scenario One

Group 1		Group 2		Group 3	
Ben	155	Tom	195	Art	130
Sam	155	Joe	195	Jon	130
Abe	160	Bif	200	Al	135
Sid	165	Dan	205	Ed	140
Les	165	Sal	205	Hal	140
$\overline{X}_1 = 160$		$\overline{X}_2 = 200$		$\overline{X}_3 = 135$	

$\overline{X}_G = 165$

Scenario Two:

Group 1		Group 2		Group 3	
Ben	95	Tom	115	Art	80
Sam	125	Joe	155	Jon	110
Abe	160	Bif	200	Al	135
Sid	195	Dan	245	Ed	170
Les	225	Sal	285	Hal	200
$\overline{X}_1 = 160$		$\overline{X}_2 = 200$		$\overline{X}_3 = 135$	

$\overline{X}_G = 165$

In both scenarios, the between groups variance is exactly the same because the group means and grand means are identical. However, the within groups variance will be much larger in the second scenario than in the first, because the individual scores are far more spread out within each group in Scenario Two. Because of the lack of consistency in scores within groups in Scenario Two it is much less likely that the observed differences in group means are due to true weight differences in the groups and more likely that these differences are due to random factors since the weights are so varied. In contrast in Scenario One, it really appears that the people in Group 2 weigh more on average than so the people in Group 3, since there is much less fluctuation within each group, but a lot of variation across the groups. So, the best scenario for detecting significant differences between groups is when the difference between groups is large (the groups differ as much as possible from each other) and the difference within groups is small (the individual scores within each group are as similar as possible).

Degrees of Freedom

As with the t-test, an F-test is conducted with some degrees of freedom. If you recall for the t-test one degree of freedom was lost for each of the two samples (and there are always only two groups with the t-test), so the degrees of freedom (or how many scores are free to vary) was equal to $df = (N_1 - 1) + (N_2 - 1)$. Remember that degrees of freedom all but the last score can vary. Overall for a study there will be $N - 1$ degrees of freedom. For the example of the study-guide group, help-session group, and control group, with 15 subjects per group (45 subjects in the entire study), 44 of the scores can vary to be anything, but in order to obtain the overall grand mean of 85, the final score is fixed. So for the entire study, there are $N - 1 = 45 - 1 = 44$ degrees of freedom. These 44 degrees of freedom are divided between the between-groups variance and the within-groups variance.

Let's consider between-groups variance first. With between-groups variance we are considering variance between groups. We always lose one degree of freedom for however many things we're looking at. For between-groups variance we are looking at groups, which in this example are three. We lose one degree of freedom, so the degrees of freedom for between groups in this particular example is two. Generally, the degrees of freedom associated with between-groups variance is equal to the number of groups you have (K) minus the one degree of freedom you lose: $df_B = K - 1$.

Now for the within-groups variance we are considering the variance of individual scores within each group. Let's take the study-guide group first. There are a total of 15 individual scores and we lose one degree of freedom, so that equals 14. For the help-session group the same thing happens: we lose one degree of freedom, so we take the 15 scores minus one equals 14. And, for the control group with 15 students the same thing: number of scores minus one = 15 - 1 = 14. So, we lost one degree of freedom in each of the three groups, and ended up with a within-groups degrees of freedom equal to: 14 + 14 + 14 = 42. You can compute degrees of freedom for within-groups variance by taking the number of scores in each individual group and subtracting one, then adding all these up: $df_W = (N_1 - 1) + (N_2 - 1) + ... + (N_K - 1)$ where K is the total number of groups you have. So, in this example, $df_W = (15 - 1) + (15 - 1) + (15 - 1) = 14 + 14 + 14 = 42$. You can simplify this formula if you wish by taking the total number of subjects in the study and subtracting the number of groups you have (another way to indicate that you lose one degree of freedom for each of the groups): $df_W = N - K = 45 - 3 = 42$.

Notice that if we add up the two sets of degrees of freedom we get 2 (for df_B) and 42 (for df_W) equals 44, which is the total degrees of freedom we said existed for the study: $df_T = N - 1$. One other interesting thing to note about degrees of freedom is that these degrees of freedom formulas really are what you used in the t-test (though you didn't know it at the time). Consider between-groups degrees of freedom for a t-test. You always have two conditions or groups with a t-test. So, for between-groups variance you lose one degree of freedom off the total number of groups: $df_B = K - 1 = 2 - 1 = 1$. There is always one degree of freedom for between groups variance with the t-test. This is traditionally just not displayed because it is always equal to one. In contrast, with the ANOVA, the number of groups you have will vary, so it is important to present the between-groups degrees of freedom. The second set of degrees of freedom associated with within-groups variance is equal to the number of subjects minus the number of groups or one degree of freedom lost for each group: $df_W = N - K = (N_1 - 1) + (N_2 - 1) + ... + (N_k - 1)$. In the two group case this formula is $df_W = (N_1 - 1) + (N_2 - 1)$, which is exactly the t-test degrees of freedom formula. What we have illustrated here is that really the t-test is nothing but a very specific case of analysis of variance with two groups. In fact, if you take the t value you compute for two groups and square it you will get the test statistic for the ANOVA (which is called the F): $F = t^2$. The t needs to be squared because we squared the differences (sum of squares) in computing the F, so the equivalent t value must also be squared (alternatively, $t = \sqrt{F}$).

Now let's look at the ANOVA table with the degrees of freedom added in:

Source	SS	df
Between	50	2
Within	300	42
Total	350	44

Calculating the F-Statistic

We are just about ready to compute the test statistic, F, for the ANOVA. The last thing we need to do involves essentially computing an average measure of between and within groups variance. Recall that when we computed sum of squares we summed up the squared differences between groups and within groups. Now what we want to do is compute something akin to an average; that is, on average how much difference is there between groups compared to on average how much difference is there within groups. Think about it this way: on average, we want the difference between groups to be very large (for the groups to be very dissimilar from each other) and we want the difference within groups to be very small (for the scores within each group to be very similar). We compute something like an average (though, technically, it is not exactly an average) by simply dividing each sum of squares by its corresponding degrees of freedom. We call this the mean squares. The **mean squares** is approximately the average or mean of the sum of squares:

$$MS = \frac{SS}{df}$$

So, for the mean squares between-groups we take the sum of squares between-groups ($SS_B = 50$) and divide by the associated degrees of freedom ($SS_W = 2$):

$$MS_B = \frac{SS_B}{df_B} = \frac{50}{2} = 25$$

Similarly, for the mean squares within we take the sum of squares within-groups ($SS_W = 300$) and divide by the corresponding degrees of freedom ($df_W = 42$):

$$MS_W = \frac{SS_W}{df_W} = \frac{300}{42} = 7.14$$

SINGLE-FACTOR ANALYSIS OF VARIANCE

Let's now take a look at the revised ANOVA table:

Source	SS	df	MS
Between	50	2	25
Within	300	42	7.14
Total	350	44	

The last thing we are left with in computing the test statistic, F (the calculated value), is to compare the mean squares between to the mean squares within. In other words, we want to come up with a single number that represents how much variance on average there is between groups compared to the amount of variance on average within groups. We do this by simply dividing the mean squares between by the mean squares within:

$$F = \frac{MS_B}{MS_W}$$

This F-value provides a measure of how much the three group means differ from one another. If there is a lot of difference between groups, mean squares between will be large and so too will F. A higher F means a greater chance of concluding a significant difference among groups (just as a higher t-value meant a greater chance of concluding a significant difference between two groups). If there is a lot of difference or variability among the scores within groups, then mean squares within will be large and F will be smaller. A smaller F means a lesser chance of concluding there is a significant difference among groups. Let's now look at the ANOVA table with the F-statistic calculated:

Source	SS	df	MS	F
Between	50	2	25	3.50
Within	300	42	7.14	
Total	250	44		

170 SINGLE-FACTOR ANALYSIS OF VARIANCE

The F-Table

So, F = 3.50 is the value we computed based on the actual 45 individual scores who were divided into three groups of 15 subjects each. We still do not know at this point whether this figure of $F = 3.50$ meaning that the groups are significantly (not due to chance) different from one another. Recall what we did with the calculated value for the t-test. A calculated value is compared to the critical value beyond which only 5% (or 1%) of all differences fall. If the calculated value is larger than the critical value, then you reject the null hypothesis and conclude that there is a significant difference between groups at some significance level (.05 or .01) with a corresponding 5% or 1% chance of being wrong (Type I error). Exactly the same thing is done with the F-test (and with all other statistical tests that we will conduct). The only thing you need to know once you have computed the test statistic is what critical value you should refer to. Well, again, recall that with the t-test you utilized the degrees of freedom (which were a function of the two sample sizes) to refer to the corresponding t-distribution that excluded the extreme outermost 5% or 1% of the distribution (using the t-table). The F-test works the same way. Use the degrees of freedom (which are a function of the sample size -- within groups, and the number of groups you are working with -- between groups) to find the critical value. The critical values for the F-test are shown in the F-table (see Appendix A). As you can see, there are two associated degrees of freedom: for between and within groups. The numerator degrees of freedom are the df associated with the between-groups variance (because MS_B is the numerator of the F calculation) and the denominator degrees of freedom are the df associated with the within-groups variance (because MSw is in the denominator of the F calculation). For each set of degrees of freedom (between and within) there are two critical values: one for the .05 significance level (the top number) and one for the .01 significance level (the bottom number).

Let's see whether the F-value we calculated ($F = 3.50$) is significant, meaning that there is a significant difference in means among the three groups. The F-value corresponding to degrees of freedom of 2 and 42 at the .05 significance level is $F_{(2,42).05} = 3.22$. Let's compare the calculated to the critical value. (Note that because the concept of two-tail versus one-tail is not relevant for analysis of variance we do not need to worry about the sign of the calculated or critical value). Since the calculated value of $F = 3.50$ is larger than the critical value at the .05 level of $F = 3.22$ we can reject H_0 at the $p < .05$ level and conclude that there is a significant difference among groups. Some of the three groups differ significantly from some other of the three groups.

With a .05 significance level, we know that there is a Type I error of .05, representing a 5% chance that we incorrectly rejected the null hypothesis and concluded there was a significant difference among groups when in fact we should have not rejected H_0 and concluded no significant difference among groups. Can we do any better than a 5% chance of making a wrong decision? Let's test the calculated F at the $p < .01$ level. The critical value for F at the .01 level from the F-table is $F_{(2,42).01} = 5.15$. Since the calculated value $F = 3.50$ is not larger than the critical $F = 5.15$, we cannot reject H_0 at the $p < .01$ level. The final decision must be to reject H_0 at the $p < .05$ level (and accept a 5% error rate since we can't reject H_0 at $p < .01$), and conclude that there is some significant difference among the three groups.

So which of the three groups differ significantly from which other groups? Did the help session cause students to score significantly higher ($\overline{X}_{HS} = 90$) than the control group ($\overline{X}_C = 80$) and the study-guide group ($\overline{X}_{SG} = 85$)? Did the help-session group score significantly higher than the control group but not the study-guide group? Did the study-guide group score significantly higher than the control group? To answer these questions about which groups differ from which other groups it is necessary to conduct something called a post-hoc test. A **post-hoc test** is a statistical test conducted after a primary statistical test (e.g., the F-test) to assess which groups differ significantly from which other groups. For the F-test, the post-hoc tests typically involves comparing two means at a time. It is sort of like conducting a t-test following the F, except that unlike the t-test, the post-hoc test controls for the initial Type I error of .05 or .01 achieved by conducting the F-test rather than the t-test in the first place. Post-hoc tests come in several different varieties, the most popular of which is the Scheffé (which allows you to compare all means to all other means). Although post-hoc tests are a bit beyond the scope of this introductory textbook, you should be aware that this is the step you conduct following a significant F in order to determine which groups differ significantly from which other groups. Notably the most popular post-hoc method you will see used is the Scheffé, since this procedure allows you to compare all group means with each other while simultaneously controlling the Type I error at .05 or .01 as achieved by the significant F.

An ANOVA (F-Test) Example

Although we did an example above as we explained each of the basic concepts of the ANOVA, it might be useful to run through another example now that you are familiar with all the procedures and all the terminology for conducting an F-test. Consider the following problem: a sociologist is interested in the effects of having regular family meetings on children's delinquent behavior. The researcher hypothesizes, based on the review of the literature in the area of children and delinquent behavior, that children who receive positive reinforcement from their family members will be less likely to exhibit delinquent behavior than will children who do not receive positive reinforcement from their families. The researcher finds 100 children who can be considered to have behavioral problems and randomly assigns 25 of the children to have one-hour meetings one time per week with their families to talk about positive achievements in the child's life. Another 25 of the children are randomly assigned to have one-hour meetings two times per week with their families. Twenty-five other children are randomly assigned to have one-hour meetings four times per week with their families. And, the final 25 children are randomly assigned to have no meetings with their families. After a period of two months the researcher goes to the children's schools and obtains ratings from teachers and peers on each child's aggressiveness, amount of trouble-making, and other delinquent types of behavior. These ratings are scaled from 0 to 10, with 0 indicating no delinquent behavior and 10 indicating a high amount of delinquent behavior. The researcher obtains the following means for the four groups in this study (the subscript on the means indicates the number of family meetings per week): $\overline{X}_0 = 9.3$; $\overline{X}_1 = 8.2$; $\overline{X}_2 = 8.0$; $\overline{X}_4 = 6.8$. The mean behavioral rating for all 100 children is $\overline{X}_G = 8.1$.

QUESTION: Does positive reinforcement by family members during regular family meeting reduce the amount of negative behavior exhibited by delinquent youths?

ANSWER:
Step 1: What are the hypotheses for this study?

As with the t-test, let's proceed by showing the steps used to solve an ANOVA problem. First, what is the researcher predicting? Well, in this case the researcher is essentially predicting that increased positive interaction with one's family will reduce the amount of children's delinquent behavior. Let's consider what the independent and dependent variables are in this study. The independent variable (which the researcher manipulates) is the frequency of positive interaction with one's family members. There are four levels to this one independent variable: 0 times per week (the control), 1 time per week, 2 times per week, and 4 times per week (you can think of these as three experimental conditions). The dependent variable is the teacher/peer ratings of the child's negative behavior on a scale of 0 to 10. Since the dependent measure is interval-level data (recall from chapter 2 that there is some debate regarding whether Likert-type scales yield ordinal- or interval-level data) and since there is one independent variable with more than two groups, we know that we will be conducting an analysis of variance, or an F-test.

Now, what is the researcher predicting? Well, he is predicting that those who receive greater positive family interaction will exhibit a lower mean delinquent behavioral score than will those who receive less positive family interaction:

$$H_a: \mu_0 > \mu_1 > \mu_2 > \mu_4 \quad \text{or} \quad H_a: \mu_4 < \mu_2 < \mu_1 < \mu_0$$

The corresponding null hypothesis is the means will be no different from one another (or in the opposite than predicted direction):

$$H_0: \mu_0 \leq \mu_1 \leq \mu_2 \leq \mu_4 \quad \text{or} \quad H_0: \mu_4 \geq \mu_2 \geq \mu_1 \geq \mu_0$$

Although this is a directional hypothesis, it is not considered one-tail because with analysis of variance the concept of one-tail and two-tail does not exist (since we are not working directly from the distribution of differences, which compares only two sample means).

Step 2: Begin the ANOVA table and compute the sum of squares.

The next step in conducting the F-test is to compute the sum of squares between groups and the sum of squares within groups. Let's compute the sum of squares between groups using the four group means ($\overline{X}_0 = 9.3$, $\overline{X}_1 = 8.2$, $\overline{X}_2 = 8.0$, $\overline{X}_4 = 6.8$) we obtained and the overall grand mean ($\overline{X}_G = 8.1$) for all 100 children in the study:

0 family meetings	:	9.3 - 8.1 = +1.2	$+1.2^2 = 1.44$
1 family meeting	:	8.2 - 8.1 = +0.1	$+0.1^2 = .01$
2 family meetings	:	8.0 - 8.1 = -0.1	$-0.1^2 = .01$
4 family meetings	:	6.8 - 8.1 = -1.3	$-1.3^2 = 1.69$
		Sum of squares between	*= 3.15*

So, the sum of squares between groups is $SS_B = 3.15$. Let's assume you have each of the 25 individual scores for children in each of the four groups and you compute the sum of squares within groups and obtain $SS_W = 98.63$. Filling in the ANOVA table so far, we have the following:

Source	SS
Between	3.15
Within	98.63
Total	101.78

Step 3: Compute the degrees of freedom.
Recall that there are two sets of degrees of freedom: one associated with the between-groups variance and one associated with the within-groups variance. For between groups you lose one degree of freedom across the total number of groups in the study: $df_B = K - 1 = 4 - 1 = 3$ degrees of freedom. There are three degrees of freedom between groups (one less than the total number of groups in the study). For within groups you lose one degree of freedom across the total number of subjects you have per group: $df_W = (N_1 - 1) + (N_2 - 1) + (N_3 - 1) + (N_4 - 1) = (25 - 1) + (25 - 1) + (25 - 1) + (25 - 1) = 24 + 24 + 24 + 24 = 96$; or, $df_W = N - K = 100 - 4 = 96$ degrees of freedom. There are 96 degrees of freedom within groups (one less than the number of subjects per group). The total number of degrees of freedom is $df_T = df_B + df_W = 3 + 96 = 99$; or $df_T = N - 1 = 100 - 1 = 99$ degrees of freedom. The total number of degrees of freedom in the entire study equals the total number of subjects participating less one degree of freedom (all but the last score can vary in order to obtain the grand mean of the study). The ANOVA table now looks like the following:

Source	SS	df
Between	3.15	3
Within	98.63	96
Total	101.78	99

Step 4: Compute the mean squares.

The mean squares are simply a value representing approximately the average distance scores are away from the mean (either group means away from the grand mean for between groups or individual scores away from group means for within groups) using the associated degrees of freedom. Recall that the general formula for MS is:

$$MS = \frac{SS}{df}$$

So, let's compute the MS for between and within groups. For between groups:

$$MS_B = \frac{SS_B}{df_B} = \frac{3.15}{3} = 1.05.$$

For within groups:

$$MS_W = \frac{SS_W}{df_W} = \frac{98.63}{96} = 0.40$$

Now the ANOVA table looks like the following:

Source	SS	df	MS
Between	3.15	3	1.05
Within	38.63	96	0.40
Total	41.78	99	

Step 5: Compute the F-value.

Recall that the F-value is computed by comparing the variance between groups to the variance within groups:

$$F = \frac{MS_B}{MS_W} = \frac{1.05}{0.40} = 2.63$$

So, the completed ANOVA table now looks like the following:

Source	SS	df	MS	F
Between	3.15	3	1.05	2.63
Within	38.63	96	0.40	
Total	41.78	99		

Step 6: Look up the critical value at the $p < .05$ level.

This ANOVA problem involves 3 and 96 degrees of freedom, and we use the F-table to look up the critical value. As it happens, there is no denominator degrees of freedom of 96 listed in the F-table. The closest values are 80 and 100:

$F_{(2,80).05} = 3.11$
$F_{(2,100).05} = 3.09$

Since 96 is pretty close to 100, we can take the critical value of 3.09 as a good estimate of the true critical value. This process is called extrapolation when you find the critical value for a certain degrees of freedom that is not listed in the F-table using the nearest listed degrees of freedom.

Step 7: Compare the calculated F to the critical F at the $p < .05$ level.

178 SINGLE-FACTOR ANALYSIS OF VARIANCE

Since the calculated F or 2.63 is not larger than the critical F of 3.09 we can stop here and not reject the null hypothesis. There is no significant difference among the four group means.

Step 8: Look up the critical value at the $p < .01$ level.

Not applicable, since we could not reject H_0 at $p < .05$. This step is irrelevant (though, by extrapolation, the critical value would be approximately 4.88 or 4.89).

Step 9: Compare the calculated F to the critical F at the $p < .01$ level.

Again, not applicable, since we could not reject H_0 at $p < .05$.

Step 10: State the final conclusion both statistically and in words.

The final conclusion stated statistically is to not reject H_0. In words, we conclude that there is no significant difference among the four groups. Positive family interaction does not cause a significant reduction in delinquent behavior among delinquent children. Note that if we had rejected H_0 we would need to have conducted a post-hoc test (such as a Scheffé) in order to determine which groups were significantly different from which other groups.

The following table provides a summary of the formulas used for completing the one-way (single-factor) ANOVA table:

		ONE-WAY ANOVA SUMMARY TABLE		
Source of Variation (Source)	Sums of Squares (SS)	Degrees of Freedom (df)	Mean Squares (MS)	Fobs (F)
Between		(K-1)*	$\frac{SS_B}{df_B}$	$F_{obs} = \frac{MS_B}{MS_W}$
Within		(N-K)*	$\frac{SS_W}{df_W}$	
Total	$SS_T = SS_B + SS_W$	N-1		

*where K=number of groups
*where N=number of subjects

Table 7.1 One-way ANOVA summary table

Chapter 7 Problems

1. If you want to compare the difference between sample means for only two groups you should conduct a(n) _____, whereas if you want to compare the differences among more than two sample means you should conduct a(n) _____.

2. What is another name for an independent variable?

3. What is the formula for an F?

4. What is the formula for computing the degrees of freedom for between-group differences?

5. What is the formula for computing the degrees of freedom for within-group differences?

6. In order to detect a significant difference among groups, which do you want larger: between-group differences or within-group differences?

7. If you find a significant F, what is your next procedure?

8. A(n) _____ test tells you that there is a significant difference among some of the group means, whereas a(n) _____ test informs you which groups in particular differ from one another.

9. Can a t-test be used as a post hoc test? Why or why not?

10. With a t you can conduct a one-tail or a two-tail test. Can you do the same with an F?

11. For each of the following, indicate what your statistical decision would be (reject H_0 at $p < .05$, reject H_0 at $p < .01$, do not reject (accept) H_0, or cannot be determined from the information given).

   a. $F = 2.91$ 4 groups   10 subjects per group
   b. $F = 2.70$ $df_B = 9$ $df_W = 80$
   c. $F = 3.00$ 5 groups   15 subjects in group 1
   d. $F = 3.67$ 3 groups   17 subjects in the study
   e. $F = 2.40$ 13 groups   32 subjects per group

12. Given the summary table below:

Source	SS	df	MS	F
Between	82.8	3	27.60	
Within	69.2	16	4.32	
Total	152.0	19		

   a. How many groups were in this study?

   b. How many subjects participated in this study?

   c. Conduct an F-test.

   d. State your decision both statistically and in words.

13. Given the summary table below:

Source	SS	df	MS	F
Between	50	5	10	
Within	25	5	5	
Total	75	10		

   a. How many groups were in this study?

   b. How many subjects participated in this study?

   c. Compute the F-value.

   d. State your decision both statistically and in words.

14. A researcher hypothesizes that children who view different types of *Star Trek* programs will have different beliefs about the possibility of life on other planets. Forty children are taken from a 5th grade classroom and randomly assigned to watch one of four different versions of *Star Trek*. An equal number of children are randomly assigned to watch the old *Star Trek* program, the new *Star Trek* program (*TNG*), a cartoon version of the old *Star Trek* program, and a cartoon version of the new *Star Trek* program. After children view the appropriate program they are asked how possible it is for life to exist on other planets on a scale ranging from 1 to 5, where 1 represents "not at all possible" and 5 represent "very possible". The data yield the following partial ANOVA table.

Source	SS	df	MS	F
Between	55			

Within	140
Total	195

$\overline{X}_{old} = 2.34$ $\overline{X}_{new} = 4.56$ $\overline{X}_{old\ cart} = 1.44$ $\overline{X}_{new\ cart} = 1.69$

a. State the null and research hypotheses.

b. How many children watched the cartoon version of the new *Star Trek* program?

c. Complete the F-table and conduct an F-test.

d. State your decision both statistically and in words.

e. Basing your answer on your analysis, which group(s) are significantly more likely to believe that life exists on other planets?

Suppose your ANOVA table was as follows:

Source	SS	df	MS	F
Between	25			
Within	85			
Total	110			

$\overline{X}_{old} = 4.44$ $\overline{X}_{new} = 4.20$ $\overline{X}_{old\ cart} = 4.35$ $\overline{X}_{new\ cart} = 4.10$

f. Complete the F-table and conduct an F-test.

g. State your decision both statistically and in words.

h. Basing your answer on your analysis, which group(s) are significantly more likely to believe that life on other planets is possible?

15. An experimenter hypothesizes that factory workers will differ in their productivity according to how much positive reinforcement they receive from their supervisors. The researcher randomly assigns 10 employees to receive, unbeknownst to them, 5 minutes of positive reinforcement per day from their immediate supervisors. Ten other employees are randomly assigned to receive 10 minutes of praise per day, and 10 more employees receive 15 minutes of praise per day. One month later the researcher returns and measures the employees' productivity rates in terms of number of Widgets produced per hour. The data obtained showed the following:

Source	SS	df	MS	F
Between	180			
Within	800			
Total	980			

$\overline{X}_{5\,min} = 836$ $\overline{X}_{10\,min} = 884$ $\overline{X}_{15\,min} = 912$

a. State the null and research hypotheses.

b. How many total employees participated in this study?

c. Complete the F-table and conduct an F-test.

d. State your decision both statistically and in words.

e. Basing your answer on your analysis, how many minutes per day would you recommend that supervisors praise their employees to obtain significantly higher productivity rates?

Instead of 10 employees per condition, suppose there were had 20 employees per condition with the remainder of the information identical to that above:

Source	SS	df	MS	F
Between	180			
Within	800			
Total	980			

$\overline{X}_{5\,min} = 836$ $\overline{X}_{10\,min} = 884$ $\overline{X}_{15\,min} = 912$

f. Complete the F-table and conduct an F-test.

g. State your decision both statistically and in words.

h. Basing your answer on your analysis, how many minutes per day would you recommend that supervisors praise their employees to obtain significantly higher productivity rates?

CHAPTER 8
MULTIPLE-FACTOR ANALYSIS OF VARIANCE (ANOVA)

Single vs. Multiple Factor ANOVA

Let's begin this chapter by reviewing what we know about statistical tests so far. To determine what statistical test is appropriate, you need to determine: 1) what level of measurement you have, and 2) how many independent variables and groups you have. All of the tests we have discussed so far involve interval- or ratio-level data in which a mean can be computed. And, all of these tests involve having one dependent variable only (e.g., exam score). The z-test was a theoretical test that assumed you have perfect knowledge of the population; the z-test involved one independent variable with two levels (two groups). The version of the z-test that is actually conducted in practice is the t-test because you almost never have perfect knowledge of a population. The t-test involves having one independent variable (also called a factor) and two levels (conditions or groups). The F-test is used any time you have more than a total of two groups in the study. Single-factor ANOVA is an F-test that involves comparing three or more groups for one independent variable. The F-test we will introduce in this chapter is the multiple-factor ANOVA which involves having more than one independent variable (factor) with at least two levels on each factor. This introduces the concept of the factorial design.

Factorial Designs

Most things that happen in the world are not determined by one and only one cause. Most of the time events are multiply determined. For example, the type of person you are today is influenced by your family, your peers, your school, your church, you life experiences, and so forth (i.e., multiple factors). Frequently, when researchers wish to analyze causal relationships by conducting experiments, they want to look at multiple causes of a particular event or occurrence. Let's consider the problem we looked at in the last chapter. The researcher was interested in the effect of a particular type of study technique (the one independent variable) on students' exam performance (the one dependent variable). The independent variable, study technique, had three levels or conditions: use of a study guide, attendance at an extra help session, and no study guide or extra help (control). Consider now a situation in which we have a total of 60 subjects to allocate to the conditions in this study. How many subjects would we assign to each condition? The answer is 20:

Independent Variable:	Type of study technique	
Level 1	*Level 2*	*Level 3*
Study Guide	Help Session	Control
$N_{SG} = 20$	$N_{HS} = 20$	$N_C = 20$

This is fine. This is a perfectly acceptable one-way (or one-factor, or one independent variable) design that could be conducted. But, we can also look at this problem as a multiple-factor study. That is, instead of considering that there is one independent variable (type of study technique), we might also consider that there are two independent variables: presence of a study guide (yes, no) and presence of an extra help session (yes, no). We can now cross these two factors into what is called a factorial design:

```
                Presence of Study Guide
                   Yes            No
              ┌─────────────┬─────────────┐
         Yes  │      I      │     II      │
              │             │             │
Presence of   │             │             │
Help Session  ├─────────────┼─────────────┤
         No   │     III     │     IV      │
              │             │             │
              └─────────────┴─────────────┘
```

Figure 8.1 A 2 x 2 factorial design with two independent variables: presence and absence of a study guide and presence and absence of a help session

Consider Block III (bottom left). This block represents a group of students who got the study guide (Yes) and did not get the help session (No). This corresponds (in part) with the first group in the single-factor design: study-guide group. Now consider Block II (top right). This block represents a group of students who did not get the study guide (No) and did get the help session (Yes). This corresponds (in part) with the second group in the single-factor design: help-session group. Now consider Block IV (bottom right). This block represents a group of students who did not get the study guide (No) and who did not get the help session (No). This corresponds with the final group in the one-way layout: the control group. So what group is there in Block I, then? Well, the students in the group in Block I represent an added bonus of a factorial design. This group of students received both the study guide (Yes) and the help session (Yes); this group represents a combination of the two factors. Do the two factors (study guide and help session) interact with one another in any way, such that perhaps the highest exam score occurs among students who *both* use the study guide and attend the help session? The ability to test for this interaction is not possible with the one-way layout. No group received both the study guide and help session in the single-factor ANOVA design. So, one added benefit of a factorial design over a single-factor layout is the ability to test for interactions of independent variables.

A second benefit of the factorial design comes in the form of increased power. Recall that power refers to the ability to detect significant differences among groups. One way to increase power that we discussed in detail in Chapter 6 (t-test) was that a larger sample size increases power. The more people you have in each group, the more ability you have to detect a significant difference among the groups. Note that in the one-way layout there was a total of 60 students; of these, 20 students received the study guide and 20 students went to the help session. Now consider if we were to assign these same 60 students to the conditions in the factorial design (where N = number of subjects):

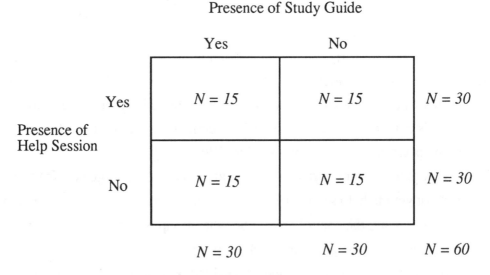

Figure 8.2 Random assignment of a total of $N = 60$ subjects to conditions of a 2 x 2 factorial design.

In this factorial design, how many total students used the review sheet? 30 students! Cover up the left hand side of the chart that says Presence of Help Session -- Yes/No. Now look down the column labeled Yes for Presence of Study Guide. The answer is not 15 students who used the study guide, the answer is a full 30 students. This is 10 students more than used the study guide in the single-factor example. More participants means a greater ability to detect significant differences! How many students attended a help session (regardless of whether they used the study guide -- students' condition on this second independent variable is irrelevant). Again, if you cover the top part of the diagram labeled Presence of Study Guide -- Yes/No, and look across for Presence of Help Session = Yes, you see that 30 students attended the help session. Again, this is 10 more than attended the help session in the one-way layout. So, to review, a factorial design involves having two or more independent variables, each with two or more levels. A factorial design allows you to examine interactions of more than one variable. A factorial design also provides increased power over a conventional one-way (single factor) layout due to an increased number of subjects on levels of the independent variables.

The example we used above was a 2 x 2 factorial design. The first number, 2, refers to the first independent variable: presence of study guide. This factor had two levels (Yes, No). The second number, 2, refers to the second independent variable: presence of help session. This second factor also had two levels (Yes, No). There were a total of 2 x 2 = 4 conditions or groups in this study (see Figures 8.1, and 8.2). This is the simplest form of a factorial design. Other factorial designs may have more than two factors (independent variables) and/or more than two levels on each independent variable. Consider the following example: a researcher is interested in whether there is a difference in heart rate among older (over 50) vs. younger (under 50) males and females who jog, swim, ride bicycle, or play tennis for one hour per day. The dependent measure is heart rate, which is ratio-level data. This study involves three independent variables: IV #1 = age group (over 50, under 50); IV #2 =

gender (male, female); IV #3 = exercise activity (jogging, swimming, biking, tennis). This is a 2 x 2 x 4 factorial design. There are three numbers because there are three independent variables. The first number = 2 because there are two levels (over 50, under 50) to the first independent variable, age group. The second number = 2 because there are two levels (male, female) to the second independent variable, gender. The third number = 4 because there are four levels (jogging, swimming, biking, tennis) to the third independent variable, exercise activity. A pictorial layout of this factorial design might look like the following:

	Under 50			Over 50	
	Male	Female		Male	Female
Jogging			Jogging		
Swimming			Swimming		
Biking			Biking		
Tennis			Tennis		

Figure 8.3 A 2 x 2 x 4 factorial design with three independent variables: age group, gender, and exercise activity

There are a total of 2 x 2 x 4 = 16 conditions or groups in this study (you can verify this by counting the total number of boxes above).

Main and Interaction Effects

Recall that with a one-way single-factor ANOVA, there was only one F-test that we conducted: is there a difference among any of the group means on this one independent variable? Now with a factorial design, there are multiple independent variables or multiple factors. With a multiple-factor F-test there is more than one set of differences to assess. Let's consider the 2 x 2 factorial design with the study guide and help session. First, did students who used the study guide score significantly different (higher) than students who did not? This is one statistical test. Second, did students who went to the help session score significantly different (higher) than students who did not? This is a second statistical test. Third, did some combination of the study guide and help session cause students to score significantly different than some other combination (interaction) of study guide and help session? This is a third statistical test. So, with a two-factor layout, there are two possible tests for each of the two independent variables, called main effects, and one possible test for the interaction of the two independent variables with each other, called an interaction effect. A **main effect** is an effect for one independent variable. An **interaction effect** is an effect for a combination of two or more independent variables. So, we need to ascertain whether: 1) there is a main effect for study-guide condition; 2) there is a main effect for help-session condition; 3) there is an interaction between study guide and help session.

Consider the 2 x 2 x 4 example with age group, gender, and exercise activity as the three independent variables. With three independent variables, there are three possible main effects: a main effect for age group, a main effect for gender, and a main effect for exercise activity. There are also numerous possible interaction effects. First, there is a possible two-way (meaning two independent variables) interaction between age group and gender. Second, there is a possible two-way interaction between age group and exercise activity. Third, there is a possible two-way interaction between gender and exercise activity. So, there are three possible two-way interaction effects that need to be tested for. Finally, there is a possible three-way interaction (involving three independent variables) among all of the factors in the study: age group, gender, and exercise activity. So, in sum, this 2 x 2 x 4 multiple-factor ANOVA would involve seven different F-tests: three main effects and four interaction effects.

Let's return now to the example with the study guide and extra help sessions. With three statistical tests to perform (two possible main effects and one possible interaction effect), we have three sets of alternative (research) hypotheses (and corresponding null hypotheses that are tested). The set of hypotheses for the test of the main effect of study guide would be:

H_a: $\mu_{SG} \neq \mu_{No\ SG}$ (or $\mu_{SG} > \mu_{No\ SG}$, for a directional prediction; let's assume the nondirectional hypotheses for these examples)

H_0: $\mu_{SG} = \mu_{No\ SG}$ (or $\mu_{SG} \leq \mu_{No\ SG}$ for a directional prediction)

The researcher is predicting that the mean for students who use the study guide will be different than the mean for students who do not use the study guide. We can examine this idea graphically by looking at the marginal means for the independent variable presence of study guide:

MULTIPLE-FACTOR ANALYSIS OF VARIANCE

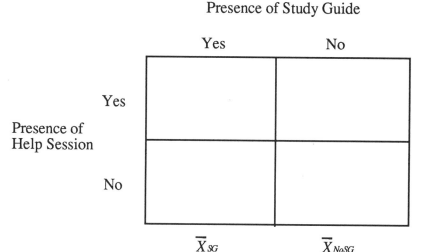

Figure 8.4 A 2 x 2 factorial design with marginal means for the two levels (presence and absence of a study guide) of the independent variable study guide

The marginal means are the ones that appear in the margins, and are the means for the levels of one independent variable (yes and no for presence of the study guide) across levels of the second independent variable. For example, the marginal mean for students who got the study guide is computed across both levels of the help session variable. It doesn't matter whether students went to the help session, the marginal mean for students who got the study guide is independent of their help session condition.

The set of hypotheses for the test of the main effect of help session would be:

$$H_a: \mu_{HS} \neq \mu_{No\ HS}$$
$$H_0: \mu_{HS} = \mu_{No\ HS}$$

Again, we can see this by looking at the difference between the marginal means for students who got the help session versus those who did not:

194 MULTIPLE-FACTOR ANALYSIS OF VARIANCE

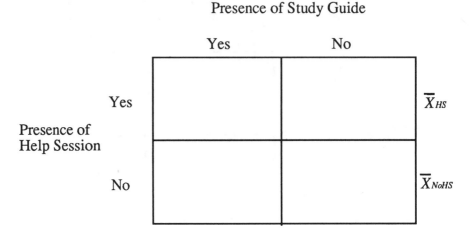

Figure 8.5 A 2 x 2 factorial design with marginal means for the two levels (presence and absence of a help session) of the independent variable help session

The marginal means for the levels of the independent variable presence of help session (yes and no) is computed across levels of the independent variable presence of study guide. The marginal mean for students who got the help session is computed across both levels of the study guide variable. It does not matter whether students received the study guide, the marginal mean for students who got the help session is independent of their study guide condition.

Finally, the set of hypotheses for the test of the interaction effect between study-guide condition and help-session condition is slightly more complex. In this case, the researcher is actually looking at the cell means (which are the means that fall in the four boxes) rather than the marginal means, because it is only the cell means that are a representation of various combinations of the two independent variables:

Presence of Study Guide

	Yes	No
Presence of Help Session — Yes	$\bar{X}_{SG-HS}$	$\bar{X}_{NoSG-HS}$
Presence of Help Session — No	$\bar{X}_{SG-NoHS}$	$\bar{X}_{NoSG-NoHS}$

Figure 8.6 A 2 x 2 factorial design with cell means for the combination of the two levels of the independent variable of study guide and help session

The marginal means represent one independent variable across levels of the other independent variable. In contrast, the cell means are the representation of which level the participant is at for each of the independent variables. For example, if you were a student in the upper left-hand box we would know that you were at the Yes level of the first independent variable (study-guide condition) and the Yes level of the second independent variable (help-session condition). In contrast, if you were a student at the marginal mean for study guide, we would know what level you were at for the study guide variable (Yes, you got the study guide), but we would have no idea about help session variable (whether you were at the Yes or No level, if you went to the help sessions or not). So, an interaction effect involves differences in cell means. For a two-way interaction these differences are between levels of one independent variable for each level of the second independent variable. For example, consider the following set of cell means:

MULTIPLE-FACTOR ANALYSIS OF VARIANCE

	Presence of Study Guide	
	Yes	No
Presence of Help Session — Yes	$\overline{X} = 90$	$\overline{X} = 95$
Presence of Help Session — No	$\overline{X} = 70$	$\overline{X} = 75$

Figure 8.7 Cell means for a 2 x 2 factorial design

Let's do this example with the first independent variable as presence of study guide with two levels: Yes and No. What is the difference in these two study-guide levels for each level of the help-session variable (the second independent variable). Well, let's consider the Yes level of the help session variable first: the difference between the two levels of the study guide variable is 90 - 95 = -5 (note that we are subtracting across, from left to right; you could also subtract from right to left if you choose). Is this difference the same as the difference between study guide levels on the No level of the help session variable? The difference between the levels of the study guide variable for the No level of the help session variable is 70 - 75 = -5. Since these two differences are the same, -5 = -5, we can conclude there is no interaction effect. That is, there is not some combination of the two variables, presence of study guide and presence of help session that differentially influenced the dependent measure final exam score.

We also could have computed differences between levels of the help-session variable for each level of the study-guide variable, and we would have obtained exactly the same results. For example, the difference between the Yes and No levels of the help-session variable for the Yes level of the study-guide variable is 90 - 70 = +20. The difference between the Yes and No levels of the help-session variable for the No level of the study-guide variable is 95 - 75 = +20. Since +20 = +20, there is no interaction effect. Note that you can subtract either across or down, from right to left or left to right, from top to bottom or bottom to top; it doesn't matter. The only thing that matters is that you do the subtraction in the same way for both levels of the second independent variable (i.e., you couldn't subtract right to left for Yes and left to right for No). Based on this examination of differences in cell means, then, we can state the research hypothesis and corresponding null hypothesis for the two-way interaction effect as follows:

H_a: $\mu_{SG/HS} - \mu_{No\ SG/HS} \neq \mu_{SG/No\ HS} - \mu_{No\ SG/No\ HS}$
H_0: $\mu_{SG/HS} - \mu_{No\ SG/HS} = \mu_{SG/No\ HS} - \mu_{No\ SG/No\ HS}$

This set of hypotheses involved subtracting the cell means in Figure 8.7 from left to right. The hypothesis for the interaction effect also could be stated in a variety of other ways depending on how you choose to subtract the means, including the following:

H_a: $\mu_{SG/HS} - \mu_{SG/No\ HS} \neq \mu_{No\ SG/HS} - \mu_{No\ SG/No\ HS}$
H_0: $\mu_{SG/HS} - \mu_{SG/No\ HS} = \mu_{No\ SG/HS} - \mu_{No\ SG/No\ HS}$

How Multiple Factor Analysis of Variance Works

How do you know, then, when a difference in marginal means is statistically significant and you have a main effect, and when a difference in cell means is statistically significant and you have an interaction effect? Well, you already know basically how to do this: conduct an F-test for each effect you ? trying to assess for significance. However, now you can have multiple ma' effects that are significant and multiple interaction effects that are significant in a single problem. In the 2 x 2 example with the study guide and the help session we know that there are two possible main effects (one for the study guide, independent variable and one for the help session independent variable) and one possible interaction effect (between the study guide and help session independent variables). The final results can be any combination of these effects: you might have two main effects and an interaction, one main effect and an interaction, two main effects and no interaction, an interaction and no main effects, one main effect only, no effects at all, and so forth.

Let's assume that we obtain the following data for this 2 x 2 problem with a total of 60 students participating:

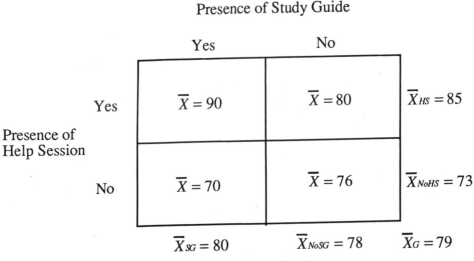

Figure 8.8 Marginal and cell means for a 2 x 2 factorial

Just as with the single-factor ANOVA, we will devise an ANOVA table that includes sum of squares, degrees of freedom, mean squares, and the test statistic, F. However, in this case, because we are conducting three statistical tests (for two main effects and an interaction effect), we will be computing three F-values. Recall from the last chapter that there were two ways in which scores could differ from one another: between groups and within groups. Now, in this case, we actually have multiple sources of between-groups variance. Scores can differ between groups on the study guide variable (the study-guide group and no-study-guide group can score differently). Scores also can differ between groups on the help-session variable (the-help session group and no-help-session group can score differently). And scores can differ between groups on the interaction of the two variables (the study-guide/help-session group can score differently than the study-guide/no-help-session group which can score differently than the no-study-guide/help-session group which can score differently than the no-study-guide/no-help-session group). So, essentially, there are three sources of between-groups variance.

And, just as there was a source of within-groups variance for the single-factor ANOVA, so too is there this same variation of individual scores within groups for the multiple-factor ANOVA. This within groups variance is slightly more complex than in the single-factor ANOVA case, because individual scores within groups can differ in each of the three ways described above. That is, individual scores can differ within the two study-guide groups; individual scores can differ within the two help-session groups; and individual scores can differ within the four groups that are combinations of the two variables. So, this within groups source of variance is fairly complex to compute, but it is analogous to how it was done in the single-factor case.

Overall, there are three between-groups sources of variance and a within groups source of variance, which is typically termed error in the multiple-factor case because whenever individual scores deviate from their group mean this is really a source of error. Recall with the single-factor ANOVA case that high within-groups variability (or error) means a smaller chance that we will be able to detect a significant difference. Recall that we want scores within each group to be as similar to each other and the group mean as possible. We can now set up a modified ANOVA table for this two-factor design as follows:

200 MULTIPLE-FACTOR ANALYSIS OF VARIANCE

Source
Study Guide
Help Session
Study Guide x Help Session
Error
Total

A problem with three independent variables would have the three independent variables as possible main effects, three possible two-way interactions, and one possible three-way interaction, so there actually would be seven sources of between-groups variance in a three-way problem, plus error (within-groups variance). In a four independent variable problem the number of sources of between-groups variance would be even larger!

Now let's do everything the same as with the single-factor ANOVA problem, except now we're computing several F's simultaneously. First, let's try to compute, for illustration purposes, the sum of squares for the study-guide main effect. Recall that the sum of squares between groups is the sum of squared differences between the group means and the grand mean:

$$\overline{X}_{SG} - \overline{X}_G = 80 - 79 = +1 \qquad +1^2 = 1$$
$$\overline{X}_{No\ SG} - \overline{X}_G = 78 - 79 = -1 \qquad -1^2 = 1$$
$$\text{Sum of squares} = 2$$

So, the sum of squares for the study-guide main effect is 2.

Let's do the same thing for the help-session group:

$$M_{HS} - M_G = 85 - 79 = +6 \qquad +6^2 = 36$$
$$M_{No\ HS} - M_G = 73 - 79 = -6 \qquad -6^2 = 36$$
$$\text{Sum of squares} = 72$$

So, the sum of squares for the help-session main effect is 72.

And we can basically do the same thing for computing the interaction effect and for the error term (using individual scores compared to group means in this latter case). We won't calculate these, but you should understand the basic idea of how these would be computed. We now have a partially filled ANOVA table that looks like the following:

Source	SS
Study Guide	2
Help Session	72
Study Guide x Help Session	211
Error	936
Total	1,221

Degrees of Freedom

Now what are the degrees of freedom for this problem. Well, you should remember that the total degrees of freedom for the entire study is the total number of subjects minus 1, so $N - 1 = 60 - 1 = 59$ degrees of freedom for the entire study. So, when you sum the degrees of freedom for the three between-group sources of variance and the error term you need to get 59 degrees of freedom. Remember what the degrees of freedom were for the between-groups variance: the number of groups there are minus one $df_B = K - 1$. Well, that's exactly how you compute the degrees of freedom for the main effects. For the study-guide variable the degrees of freedom is the number of groups minus one, $K_{SG} - 1 = 2 - 1 = 1$. For the help-session variable, the degrees of freedom is the number of groups minus one, $K_{HS} - 1 = 2 - 1 = 1$. For the interaction effect, you are looking at both variables (study guide and help session) simultaneously, so to compute the degrees of freedom for this interaction, study guide x help session, you multiply the degrees of freedom

202 MULTIPLE-FACTOR ANALYSIS OF VARIANCE

from the associated independent variables: $(K_{SG} - 1) \times (K_{HS} - 1) = (2 - 1) \times (2 - 1) = 1 \times 1 = 1$.

Finally, for the error term, in the single-factor ANOVA case the error term (within-groups variance term) was equal to the total number of subjects minus the total number of groups. The error term is computed exactly the same way in the multiple-factor case. The total number of groups in this study is 2 x 2 = 4 (see Figure 8.8). So the total degrees of freedom for the error term can be represented by $N - K_1 K_2 ... K_k$. For the example we are using this is, $N - K_{SG} K_{HS} = 60 - (2)(2) = 60 - 4 = 56$. Do we get the 59 total degrees of freedom for this study? Yes: 1 (study guide) + 1 (help session) + 1 (study guide x help session) + 56 (error term) = 59 (total). The ANOVA table now looks like the following:

Source	SS	df
Study Guide	2	1
Help Session	72	1
Study Guide x Help Session	211	1
Error	936	56
Total	1,221	59

Calculating the F-Statistic

The mean squares is computed identical to the single-factor ANOVA using the sum of squares and the associated degrees of freedom for each term:

$$MS = \frac{SS}{df}$$

So, for the study-guide main effect, the mean squares between is:

$$MS_{SG} = \frac{SS_{SG}}{df_{SG}} = \frac{2}{1} = 2$$

For the help-session main effect, the mean squares between is:

$$MS_{HS} = \frac{SS_{HS}}{df_{HS}} = \frac{72}{1} = 72$$

For the study guide x help session interaction effect, the mean squares between is:

$$MS_{SGxHS} = \frac{SS_{SGxHS}}{df_{SGxHS}} = \frac{211}{1} = 211$$

For the error term, the mean squares error is:

$$MS_{Error} = \frac{SS_{Error}}{df_{Error}} = \frac{936}{56} = 16.7$$

The ANOVA table now looks like the following:

Source	SS	df	MS
Study Guide	2	1	2
Help Session	72	1	72
Study Guide x Help Session	211	1	211
Error	936	56	16.7
Total	1,221	59	

Now, we can compute three F-values, corresponding to the three statistical tests we want to conduct. Again, as with the single-factor F-test, the

F-value is equal to the mean squares between groups divided by the mean squares within groups (error):

$$F = \frac{MS_B}{MS_E}$$

For the study-guide main effect, the F-value is:

$$F = \frac{MS_{SG}}{MS_E} = \frac{2}{16.7} = 0.12$$

For the help-session main effect, the F-value is:

$$F = \frac{MS_{HS}}{MS_E} = \frac{72}{16.7} = 4.31$$

And, for the study-guide x help-session interaction effect, the F-value is:

$$F = \frac{MS_{SGxHS}}{MS_E} = \frac{211}{16.7} = 12.63$$

The completed ANOVA table now looks like the following:

Source	SS	df	MS	F
Study Guide	2	1	2	0.12
Help Session	72	1	72	4.31
Study Guide x Help Session	211	1	211	12.63
Error	936	56	16.7	
Total	1,221	59		

We now have three F-values corresponding to the three effects we are trying to assess for significance. How do we know if we have any significant effects? Well, you already know how to do this because you've been doing this in the last three chapters. Compare the calculated value to the critical value. If the calculated value exceeds the critical value at the $p < .05$ or $p < .01$ level, then you can reject the null hypothesis and conclude a significant difference. What critical values do you use for the F-test? Those in the F-table corresponding to the between- and within-groups degrees of freedom used to calculated the F-value. We have three tests to perform.

Let's start with the main effect for study guide. Did students who used the study guide score significantly different ($\overline{X}_{SG} = 80$) than students who did not use the study guide ($\overline{X}_{No\ SG} = 78$), or is this observed difference purely due to chance? Well, we are performing a test with 1 degrees of freedom in the numerator (corresponding to the between-groups degrees of freedom) and 56 degrees of freedom in the denominator (corresponding to the error degrees of freedom). According to the F-table, the critical F at the $p < .05$ level with 1 and 56 degrees of freedom is $F_{(1,56).05} = 4.02$. Now compare the calculated F to the critical F. Since the calculated value $F = 0.12$ is not larger than the critical value $F = 4.02$ we cannot reject H_0 and must conclude there is no difference in final exam score between those who used the study guide and those who did not. That is, the study guide did not help students perform any better on the final exam compared to those students who did not used the study guide.

Now let's assess whether there is a main effect for help session. The calculated value was $F = 4.31$. The critical F has degrees of freedom of 1 and 56, so the critical F at the .05 significance level is $F_{(1,56).05} = 4.02$. Since the calculated value $F = 4.31$ is larger than the critical value $F = 4.02$, we can reject the null hypothesis at the $p < .05$ level, and conclude that there is a significant difference in final exam scores between those who went to extra help sessions ($\overline{X}_{HS} = 85$) and those who did not go to extra help sessions ($\overline{X}_{No\ HS} = 73$). So, help sessions caused students to perform significantly better on the final exam, with a Type I error of .05 (there is a 5% chance that this is not true, and that we incorrectly rejected the null hypothesis). Can we do any better than a 5% chance of being wrong? Let's look at the critical value for F at the .01 level, $F_{(1,56).01} = 7.12$. Since the calculated value F = 4.31 is not larger than the critical value F = 7.12 at the .01 significance level, we cannot reject H_0 at the $p < .01$ level. We must end up rejecting H_0 at the $p < .05$ level and live with the 5% error rate.

Finally, let's conduct the significance test for the interaction between study guide and help session. The critical F for 1 and 56 degrees of freedom (again, as it happens) at the $p < .05$ significance level is $F_{(1,56).05} = 4.02$. Since the calculated value F = 12.63 is larger than the critical value F = 4.02 at the .05 significance level, we can reject H_0 at the $p < .05$ level and conclude that there is a significant interaction effect. We currently have a Type I error of .05. Can we do any better than a 5% chance of being wrong? Well, the critical value at the .01 level is $F_{(1,56).01} = 7.12$. Since the calculated value F = 12.63 is larger than the critical value F = 7.12 we can reject H_0 at the $p < .01$ level and conclude that there is a significant interaction effect with only a 1% chance of being wrong. There is some difference in final exam scores for combinations of study guide and help session. To know exactly what these differences are (since there are four groups involved) we would need to conduct a post-hoc test such as the Scheffé, which we discussed in the last chapter.

The final conclusion then is that there is a main effect for help session and an interaction effect (and no main effect for study guide). We rejected the null hypothesis for the main effect for help session at the $p < .05$ level and we rejected the null hypothesis for the interaction effect at the $p < .01$ level.

An Example of Multiple-Factor ANOVA

Now let's do another example of multiple-factor ANOVA reviewing all of the steps. Consider the one-way layout in the previous chapter, where we had one independent variable (type of study technique) that had three levels: study guide, help session, control. Let's use this one independent variable again in the following problem. Suppose the researcher also believes that students who take vitamins will perform better on exams than students who don't take vitamins. We again have 60 subjects. The researcher randomly assigns the 60 subjects to study technique conditions (20 students per group: study guide, help session, control) and to vitamin conditions (30 students per group: yes -- vitamins; no -- no vitamins). The assignment of subjects to this 3 x 2 factorial design with a total of 3 x 2 = 6 groups or conditions now looks like the following:

	Study Guide	Help Session	Control	
Vitamins	N=10	N=10	N=10	N=30
No Vitamins	N=10	N=10	N=10	N=30
	N=20	N=20	N=20	N=60

Figure 8.9 A 3 x 2 factorial design with two independent variables: type of study technique, presence of vitamins.

This study involves two independent variables: type of study technique and presence of vitamins. It this 3 x 2 factorial design, the first independent variable, type of study technique, has three levels (study guide, help session, control), and the second independent variable, presence of vitamins, has two levels (yes, no). The dependent variable in this study again is final exam score, which is ratio-level data. Since we have interval- or ratio-level data, one dependent variable, and more than one independent variable (each with two or more levels) we know that we will be conducting a multiple-factor ANOVA. The following means are obtained for this study:

	Study Guide	Help Session	Control	
Vitamins	$\bar{X}=96$	$\bar{X}=90$	$\bar{X}=90$	$\bar{X}=92$
No Vitamins	$\bar{X}=88$	$\bar{X}=78$	$\bar{X}=74$	$\bar{X}=80$
	$\bar{X}=92$	$\bar{X}=84$	$\bar{X}=82$	$\bar{X}=86$

Figure 8.10 Marginal and cell means for a 3 x 2 factorial design

QUESTION: There are actually three questions we are asking here. Since there are two factors (independent variables), there are two possible main effects: one for type of study technique and one for presence of vitamins. There is also one possible two-way interaction effect between type of study technique and presence of vitamins.

> **Q #1:** Does the type of study technique a student uses cause a difference in how s/he performs on the final exam?
>
> **Q #2:** Does taking vitamins cause students to perform differently on the final exam compared to students who do not take vitamins?
>
> **Q #3:** Does some combination of type of study technique and presence of vitamins cause a difference in how students score on the final exam?

ANSWER:

Step 1: Identify the hypotheses for this study.

Just as there are three questions we are interested in answering, there are also three sets of hypotheses. First, consider the potential main effect for factor one, the independent variable, type of study technique. The researcher predicts that there will be some difference in final exam scores (means) among the three groups who receive a study guide, extra help sessions, or nothing (the control group):

H_a: $\mu_{SG} \neq \mu_{HS} \neq \mu_C$
H_0: $\mu_{SG} = \mu_{HS} = \mu_C$

Second, consider the potential main effect for factor two, the independent variable, presence of vitamins. The researcher predicts that there will be some difference between the two groups who either take vitamins or do not:

H_a: $\mu_{Vit} \neq \mu_{No\ Vit}$
H_0: $\mu_{Vit} = \mu_{No\ Vit}$

Finally, consider the potential interaction between factor one and factor two, that is, the combination of levels of type of study technique and presence of vitamins. The researcher predicts that some combination of these two factors will cause the difference between groups, whereby the groups are defined according to both the type of study technique used and the presence of vitamins. For example, it may be that students who take vitamins *and* use a study guide or go to extra help sessions score higher than everyone else. It is the interaction between both taking vitamins *and* using a study guide or going to an extra help session that causes students to score higher. Note that this is only one of many possibilities that exists if there is a significant interaction effect. A post-hoc test must be conducted to determine exactly which groups differ from which other groups when a significant interaction is found (or a significant main effect in which there are more than two groups).

Because this is a 3 x 2 factorial design, the hypothesis for the interaction effect is a bit more difficult to illustrate symbolically than the hypothesis in a 2 x 2 situation, so it will not be presented here. However, you should recognize by now that interactions are stated as differences between means. Suppose instead that this was a 2 x 2 situation with only two levels for factor one: study guide versus control (this was the old t-test situation for one independent variable). The researcher predicts that the difference between the vitamin and no-vitamin means for the study-guide groups will be different than the difference between the vitamin and no-vitamin means for the control groups. In this case, the null and alternative hypotheses would be as follows:

H_a: $\mu_{SG/Vit} - \mu_{SG/No\ Vit} \neq \mu_{C/Vit} - \mu_{C/No\ Vit}$
H_0: $\mu_{SG/Vit} - \mu_{SG/No\ Vit} = \mu_{C/Vit} - \mu_{C/No\ Vit}$

Step 2: Look at the partially filled ANOVA table with the sum of squares computed.

Source	SS
Study Technique	30
Vitamins	25
Study Technique x Vitamins	60
Error	275
Total	390

Recall from above that SS_{ST}, SS_{Vit}, and $SS_{ST \times Vit}$ are very similar to $SS_{Between}$ in the single-factor ANOVA. For example, the SS_{ST} is computed by taking the group means for the three study technique conditions ($\overline{X}_{ST} = 92$, $\overline{X}_{HS} = 84$, $\overline{X}_C = 82$) and comparing them to the grand mean ($\overline{X}_G = 86$). For both main effects and the interaction effect a similar process is followed in examining these between-group differences. Recall, too, that the error term is equivalent to the SS_{Within} for the single-factor ANOVA. And, SS_{Total} equals the sum of all the between group sum of squares and the error sum of squares.

Step 3: Calculate the degrees of freedom.

Remember for the main effects that the degrees of freedom are the number of groups minus one, K - 1, and for interaction effects, the degrees of freedom are the multiplication of the number of degrees of freedom of the associated factors (independent variables), $(K_1 - 1)(K_2 - 1) \dots (K_N - 1)$ So, for the factor, type of study technique:

$$df_{ST} = K_{ST} - 1 = 3 - 1 = 2$$

For the factor presence of vitamins:

$$df_{Vit} = K_{Vit} - 1 = 2 - 1 = 1$$

For the interaction of study technique with presence of vitamins:

$$df_{ST \times Vit} = (K_{ST} - 1) \times (K_{Vit} - 1) = (3 - 1) \times (2 - 1) = 2 \times 1 = 2$$

The error term is the number of subjects in the study minus the total number of groups or conditions:

$$df_E = N - (K_{ST})(K_{Vit}) = 60 - (3)(2) = 60 - 6 = 54$$

The sum of all the between and error degrees of freedom should equal the total degrees of freedom in the study:

$$df_T = N - 1 = 60 - 1 = 59$$
$$df_T = 2 + 1 + 2 + 54 = 59$$

The revised ANOVA table now looks like the following:

Source	SS	df
Study Technique	30	2
Vitamins	25	1
Study Technique x Vitamins	60	2
Error	275	54
Total	390	59

Step 4: Compute the values for the mean squares.

Using the formula for the mean squares, we can compute the mean squares for each of the between groups terms and the error term. For the study technique term:

$$MS_{ST} = \frac{SS_{ST}}{df_{ST}} = \frac{30}{2} = 15$$

For the vitamins term:

$$MS_{Vit} = \frac{SS_{Vit}}{df_{Vit}} = \frac{25}{1} = 25$$

For the interaction term (study technique x vitamins):

$$MS_{STxVit} = \frac{SS_{STxVi}}{df_{STxVi}} = \frac{60}{2} = 30$$

And, for the error term:

$$MS_{Error} = \frac{SS_{Error}}{df_{Error}} = \frac{275}{54} = 5.09$$

The revised ANOVA table now looks like the following:

Source	SS	df	MS
Study Technique	30	2	15
Vitamins	25	1	25
Study Technique x Vitamins	60	2	30
Error	275	54	5.09
Total	390	59	

Step 5: Compute the F-values.

Using the formula for the F-value, we can compute the following three F-values for the tests of the two main effects and the interaction effect. For the main effect for study technique:

$$F_{ST} = \frac{MS_{ST}}{MS_{Error}} = \frac{15}{5.09} = 2.95$$

For the main effect for presence of vitamins:

$$F_{Vit} = \frac{MS_{Vit}}{MS_{Error}} = \frac{25}{5.09} = 4.91$$

And for the interaction effect of study technique x presence of vitamins:

$$F_{STxVit} = \frac{MS_{STxVit}}{MS_{Error}} = \frac{30}{5.09} = 5.89$$

The completed ANOVA table looks like the following:

Source	SS	df	MS	F
Study Technique	30	2	15	2.95
Vitamins	25	1	25	4.91
Study Technique x Vitamins	60	2	30	5.89
Error	275	54	5.09	
Total	390	59		

Step 6: Look up the critical values and conduct the first test.

The first test is of the main effect for study technique. The critical value has 2 degrees of freedom in the numerator (for between subjects) and 54 degrees of freedom in the denominator (for the error term):

$$F_{(2,54).05} = 3.17$$
$$F_{(2,54).01} = 5.01$$

Starting with the .05 significance level, compare the calculated F to the critical F. Since the calculated value $F = 2.95$ is not larger than the critical value $F = 3.17$ at the .05 significance level we do not reject H_0 and conclude there is no difference in final exam scores as a function of type of study technique used. It is not necessary to conduct this test at the $p < .01$ significance level since the null hypothesis has already been accepted.

Step 7: Look up the critical values and conduct the second test.

The second test is of the main effect for presence of vitamins. The critical values will be using 1 degree of freedom in the numerator and 54 degrees of freedom in the denominator:

$$F_{(1,54).05} = 4.02$$
$$F_{(1,54).01} = 7.12$$

Starting with the .05 significance level, compare the calculated F to the critical F. Since the calculated value $F = 4.91$ is larger than the critical value $F = 4.02$ at the .05 significance level we can reject H_0 and conclude there is a significant difference in final exam score as a function of taking vitamins. We reach this conclusion with a 5% chance of being wrong. So, to determine if we can obtain a lower error rate we conduct this statistical test at the $p < .01$ significance level. Since the calculated value $F = 4.91$ is not larger than the critical value $F = 7.12$ at the $p < .01$ significance level we cannot reject H_0 at $p < .01$. The final conclusion is to reject H_0 at $p < .05$ and conclude there is a significant main effect for presence of vitamins with a Type I error of .05.

Step 8: Look up the critical values and conduct the next test.

The final test is for the interaction effect for study technique x presence of vitamins. The critical values have 2 degrees of freedom in the numerator and 54 degrees of freedom in the denominator:

$$F_{(2,54).05} = 3.17$$
$$F_{(2,54).01} = 5.01$$

Starting with the .05 significance level, compare the calculated F to the critical F. Since the calculated value $F = 5.89$ is larger than the critical value $F = 3.17$ at the .05 significance level we can reject H_0 and conclude there is a significant difference in final exam score as a function of some combination of type of study technique used and presence of vitamins. We reach this conclusion with a 5% chance of being wrong. To determine if we can have a lower error rate we conduct this statistical test at the $p < .01$ significance level. Since the calculated value $F = 5.89$ is larger than the critical value $F = 5.01$ at the $p < .01$ significance level we can reject H_0 at $p < .01$. The final conclusion is to reject H_0 at $p < .01$ and conclude there is a significant interaction effect for study technique x presence of vitamins with a Type I error of .01. If there were more effects to test you would repeat this step for each effect you were assessing for significance.

Step 9: State final conclusion both statistically and in words.

Statistically, the final conclusion is that we do not reject H_0 for the test of the main effect of study technique, reject H_0 at $p < .05$ for the test of the main effect of presence of vitamins, and reject H_0 at $p < .01$ for the test of the interaction effect of study technique x presence of vitamins. In words we conclude that first, there is a main effect for presence of vitamins and an interaction effect. Students who took vitamins scored significantly different on the final exam than did students who did not take vitamins. Second, some combination of study technique and presence of vitamins caused students to score differently than some other combination of type of study technique and presence of vitamins. The specific nature of this latter difference for the interaction effect can be determined only by conducting a post-hoc test.

The following table provides a summary of the formulas used for completing the two-way (multiple-factor) ANOVA table:

MULTIPLE-FACTOR ANALYSIS OF VARIANCE 217

TWO-WAY ANOVA SUMMARY TABLE

Source of Variation (Source)	Sums of Squares (SS)	Degrees of Freedom (DF)	Mean Squares (MS)	Fobs (F)
Factor A	main effect	(K_A-1)*	$\dfrac{SS_A}{df_A}$	$F_A = \dfrac{MS_A}{MS_W}$
Factor B	main effect	(N_B-1)*	$\dfrac{SS_B}{df_B}$	$F_B = \dfrac{MS_B}{MS_W}$
A x B	interaction effect	$(K_A-1) \times (N_B-1)$	$\dfrac{SS_{A \times B}}{df_{A \times B}}$	$F_{A \times B} = \dfrac{MS_{A \times B}}{MS_W}$
Within-group (error)		$N - K_A K_B$*	$\dfrac{SS_W}{df_W}$	
Total		$N-1$		

*where K_A = number of levels for factor A
*where K_B = number of levels for factor B
*where N = number of subjects

Table 8.1 Two-way ANOVA summary table

CHAPTER 8 PROBLEMS

1. If you want to compare the difference between two or more independent variables (factors) with two or more levels each, what test should you conduct?

2. For each of the following designs, indicate the number of factors (independent variables):
   a. 2 x 3
   b. 3 x 6 x 8 x 3 x 2
   c. 3 x 4
   d. 6 x 2 x 2
   e. 3 x 2 x 3
   f. 4 x 3 x 2 x 5

3. For each of the following designs, indicate the number of levels for the emphasized factor:
   a. 3 x 2 x 4   2nd factor
   b. 2 x 6 x 9 x 3 x 4   3rd factor
   c. 3 x 5 x 4 x 4   1st factor
   d. 2 x 3   2nd factor
   e. 5 x 6 x 3   4th factor
   f. 3 x 2   2nd factor

4. For each of the following designs, indicate the total number of experimental conditions (groups or cells) in the study:
   a. 4 x 3
   b. 2 x 2 x 2
   c. 2 x 3 x 4

5. For each of following the designs, indicate how many subjects participated in the study:
   a. 3 x 2 x 3   N = 5 subjects per condition
   b. 4 x 3   N = 16 subjects per condition
   c. 3 x 3 x 3   N = 20 subjects per condition

6. When you assess the impact of one factor on the dependent variable you are examining a(n) (*main* or *interaction*) effect?

7. When you assess the impact of a combination of factors on the dependent variable you are examining a(n) (*main* or *interaction*) effect?

8. A corporation wishes to determine whether a face-to-face conference or a tele-conference is viewed as more satisfactory by their national versus international firms and by their management staff versus regular employees. Satisfaction is measured on a Likert-type scale from 1 - 7, with 1 indicating the least satisfaction and 7 indicating the most satisfaction.
   a. What type of factorial design is this?
   b. How many possible main effects are there?
   c. How many possible interaction effects are there?

9. Sixty children from kindergarten, 60 children from 3rd grade, and 60 children from 5th grade are randomly assigned within grade to view a 30 minute violent program, with either 0, 5, or 10 acts of violence per minute. Following this, within grade, children are again randomly assigned to do one of the following activities: read a short book, play on a swingset, watch a cartoon on TV, or sit alone in a room. The experimenter then lets each child play on the playground with his or her

classmates and observes how aggressive the child is in terms of the number of hits and kicks the child delivers to playmates.

a. What are the independent variables?

b. What are the dependent variables?

c. What type of design is this?

d. What type of analysis would you use?

e. How many children watched 5 acts of violence?

f. How many 5th graders sat alone in a room after viewing?

g. If the analysis revealed that the most aggressive children were those who viewed 10 acts of violence and then sat alone in a room, what type of effect would this be (*main* or *interaction*) and what factor(s) is/are involved?

h. or If 5th graders were more aggressive than either kindergartners 3rd graders, what type of effect would this be (*main* or *interaction*) and what factor(s) is/are involved?

10. A researcher wishes to determine if married men are more or less likely to be depressed than single men after viewing a program about either divorce or marriage. Twenty-four married men and 36 single men volunteered for the study. Half of the married men are randomly assigned to watch a program about divorce and the other half of the married men are randomly assigned to watch a program about

marriage. Similarly, the researcher assigns half of the single men to watch a program about divorce and half to watch a program about marriage. Following viewing each man is asked to indicate how depressed he feels on a scale from 1, "not at all depressed" to 7, "extremely depressed". Data obtained from this study are as follows:

Source	SS	df	MS	F
Marital Status	8.2			
Program Type	7.5			
Marital x Prog.	14.7			
Error				
Total	142.4			

$\bar{X}_{\text{Marr, Div}} = 2.04$ $\bar{X}_{\text{Marr, Marr}} = 5.76$ $\bar{X}_{\text{Sing, Div}} = 5.00$ $\bar{X}_{\text{Sing, Marr}} = 4.89$

a. State the null and research hypotheses for both main effects and the interaction effect.

b. How many married men saw the program about marriage?

c. Complete the F-table and conduct an F-test.

d. Are there any main effects or an interaction effect?

e. State your decisions in part d both statistically and in words.

Suppose, instead, that 36 married men and 36 single men had participated in this study and the ANOVA table looked as follows:

Source	SS	df	MS	F

Marital Status	
Program Type	12.6
Marital x Prog.	9.8
Error	204
Total	237.2

$\overline{X}_{\text{Marr, Div}} = 2.04 \quad \overline{X}_{\text{Marr, Marr}} = 5.76 \quad \overline{X}_{\text{Sing, Div}} = 5.00 \quad \overline{X}_{\text{Sing, Marr}} = 4.89$

f. Complete the F-table and conduct an F-test.

g. Are there any main effects or an interaction effect?

h. State your decisions in part g both statistically and in words.

11. A weight watchers program wishes to determine whether there is a difference in men's and women's ability to exert self-control with respect to consuming non-nutritional food depending on how much exposure they have to it and whether or not they are warned of the negative consequences of eating it. Sixty male and sixty female subjects who are concerned with watching their weight are randomly assigned to spend 5, 10, or 15 minutes in a room with a table full of desserts. Each subject is told not to eat the food in the room. Half of the subjects are randomly assigned to receive an additional reminder about how bad consumption of the desserts would be while the other subjects receive no warning. They are than watched from a two-way mirror and the quantity of food eaten is recorded. The following ANOVA table is obtained:

Source	SS	df	MS	F
Gender	40			

Exposure	30
Warning	20
Gend. x Expos.	60
Gend. x Warn.	20
Expos. x Warn.	20
Ge. x Ex. x Wa.	80
Error	648
Total	918

a. What type of design is this?

b. How many potential main effects are there?

c. How many potential interaction effects are there?

d. Complete the F-table and conduct an F-test.

e. Are there any main effects or interaction effects?

f. State your decisions in part e both statistically and in words.

CHAPTER 9
CORRELATION

What is Simple Correlation?

This chapter represents a kind of breather from the heavier work involved in conducting the F-test. Correlation is really a very simple concept. In fact, you already know what correlation is and you probably use this concept frequently. For example, based on your experiences in school, you might assume that there is some kind of positive or direct relationship between the amount of time you spend studying and your performance on exams: the more time you study the higher your score; the less time you study the lower your score. This positive, direct relationship can be graphed as an upward sloping line:

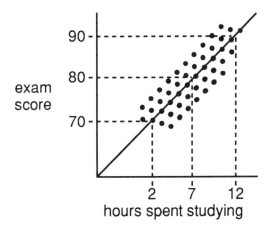

Figure 9.1 Graph of a positive relationship

Note that for 7 hours of studying the corresponding exam score is 80. If the amount of time spent studying increases to 12 hours the corresponding exam score increases to 90. If the amount of time spent studying decreases to 2 hours the corresponding exam score decreases to 70.

Similarly, you might assume that there is a negative or inverse relationship between the amount of partying you do and your grades: the more time you party the poorer your grades; the less time you party the higher your grades. This negative, inverse relationship can be graphed as a downward sloping line:

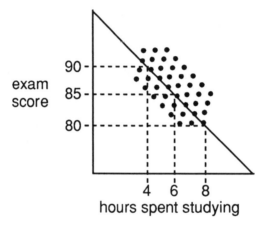

Figure 9.2 Graph of a negative relationship

If the amount of partying per week is 6 hours the corresponding exam score is 85. If the amount of time spent partying increases to 8 hours the corresponding exam score decreases to 80. If the amount of time spent partying decreases to 4 hours the corresponding exam score increases to 90.

These two examples illustrate correlations. A **simple correlation** is the relationship between two variables in terms of how these variables are related to or associated with one another; in other words, how they co-vary together. A correlation between two variables is expressed as a single number ranging between -1.00 and +1.00 and is symbolized by the statistic r. The value of a correlation has two major characteristics: 1) direction, and 2) magnitude. Direction we have already discussed above. **Direction** simply refers to whether the correlation is positive (direct) or negative (inverse), and is indicated by the sign (+ or -) of the correlation value. A **positive** or **direct relationship** means that as X increases Y increases and as X decreases Y decreases (e.g., the relationship between type spent studying and exam score). A **negative** or **inverse relationship** means that as X increases Y decreases and as X decreases Y increases (e.g., the relationship between time spent partying and exam score).

The second major component of correlation is the magnitude. The **magnitude** refers to the strength of the relationship between the variables, such that a larger number in absolute terms (ignoring the + or - sign) means that the relationship is stronger. Basically, not all variables have a perfect relationship. For example, consider the relationship between SAT scores and college GPA. A variety of educational research shows that these two variables are somewhat positively related, but not perfectly and not even very strongly. That is, someone who has a 1500 SAT score presumably will have a high college GPA and someone who has a 900 SAT score presumably will have a lower college GPA. But, sometimes students with low SAT scores have high college GPAs and vice versa, so the positive relationship between SAT scores and college GPA does not hold true all the time. In fact, the relationship might look something like the following:

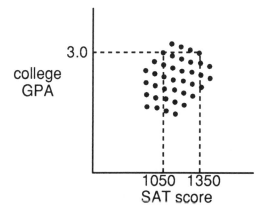

Figure 9.3 A moderate positive relationship

228 CORRELATION

Each point represents an individual, with a corresponding SAT score and college GPA. As you can see, the general trend is positive. Higher SAT scores tend to be associated with higher GPAs. But, there is a lot of variation around this. Look at the two people indicated by the two arrows. Both people have a college GPA of 3.0. However, the right-hand dot indicates that person had a pretty high SAT score (1350) whereas the left-hand dot indicates that person had a relatively lower SAT score (1050). So, there is not a perfect relationship between SAT score and GPA else all people with a 3.0 GPA would have exactly the same SAT score.

If there is a perfect relationship between two variables, then the correlation will equal either +1.00 or -1.00:

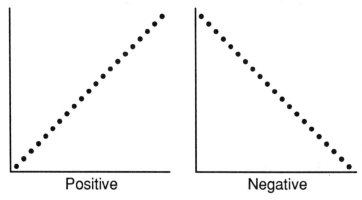

Figure 9.4 A perfect positive relationship (left) and a perfect negative relationship (right)

When the relationship between two variables is less than perfect, the correlation will be less than 1.00 (either positive or negative). We can draw a line through all of the data points and try to make the line fit the best it possibly can (we will talk more about this line in the next chapter on regression). So, for moderately correlated variables such as SAT score and college GPA in Figure 9.3 above, the line that best fits these data points might look like the following:

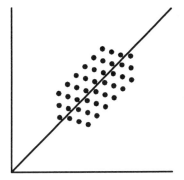

Figure 9.5 A moderate positive relationship with a line of best fit

The correlation for these data is moderately high, say $r = +.60$. The data points don't deviate too far away from this line of best fit.

Let's look at another example of the correlation between two variables, this time two that are inversely related and only very slightly correlated:

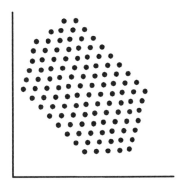

Figure 9.6 A slightly negative relationship

230 CORRELATION

If we draw the line of best fit for these data we obtain the following:

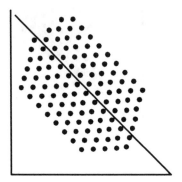

Figure 9.7 A slightly negative relationship with a line of best fit

In this case the correlation might be $r = -.20$, since there is much more variability or variation in the scores around this line. Note that the sign is irrelevant for judging the strength of the correlation: .60 is larger in terms of magnitude than .20 (regardless of the sign), so the correlation is stronger in the first example (see, Figure 9.5) than in the second (see Figure 9.7).

Now suppose that there is no relationship whatsoever between two variables: number of hairs on people's heads and the amount of TV they watch per day. These data might look like the following:

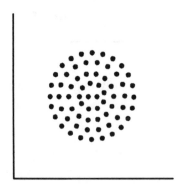

Figure 9.8 No relationship

Now, with a line of best fit we obtain the following:

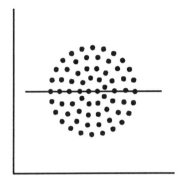

Figure 9.9 No relationship with a horizontal line

As you can see the line is horizontal. The correlation or relationship between these two variables is zero; there is no line that fits these data better than any other line, so we simply draw a horizontal line through the middle.

Let's try a practice problem. Which is stronger (has the greater magnitude): a correlation of $r = -.70$ or a correlation of $r = +.35$. The answer is $r = -.70$. Recall that magnitude refers only to the strength of the correlation (how well the data points fit around the line), not whether the relationship is positive (an upward sloping line) or negative (a downward sloping line).

Computing a Correlation, r

A simple correlation, r, is computed between two variables when the data are in interval- or ratio-level form. As with the t- and F-tests, simple correlation assumes that both sets of data are either interval- or ratio-level (we will talk about an exception to this later). So, consider computing a correlation between women's height and weight, using the following sample of data points for 10 women:

		Height (in inches)	Weight (in pounds)
1.	Mary	65	130
2.	Pat	68	150
3.	Susan	62	108
4.	Lisa	68	143
5.	Missy	66	130
6.	Angie	64	115
7.	Ariel	70	155
8.	Gwen	67	150
9.	Carol	63	120
10.	Erin	66	137

Let's order these data according to height (from shortest to tallest):

		Height (in inches)	Weight (in pounds)
1.	Susan	62	108
2.	Carol	63	120
3.	Angie	64	115
4.	Mary	65	130
5.	Erin	66	137
6.	Missy	66	130
7.	Gwen	67	150
8.	Pat	68	150
9.	Lisa	68	143
10.	Ariel	70	155

It looks as though there is a moderately strong positive relationship between height and weight: larger heights tend to have larger weights associated with them. However, you can tell this relationship is not perfect, because a taller person (Lisa at 5'8") weighs less (143 pounds) than a shorter person (Gwen at 5'7" weighs 150 pounds). So the relationship is not perfect, but weight and

properties of r:
1. $-1 \leq r \leq 1$
2. measures only strength of linear relationship
3. direction and magnitude.

height appear to be somewhat positively correlated. We won't go into the exact formula used to compute the value of r, the correlation coefficient, but it involves comparing each score to the mean of that variable (comparing each weight to the mean weight, each height to the mean height) and comparing the deviations for each person across the two variables. This r-value, computed using two sets of sample data (pairs of scores) using interval-level data is called a **Pearson product moment coefficient**, or Pearson r for short (the r for interval-level data is named after the man who first formulated how to compute it, Karl Pearson).

Testing a Correlation for Statistical Significance

Suppose we find that the correlation between height and weight is $r = +.60$. Wow, this looks pretty big. So, how big is big enough to know that this is not a random or chance value. For instance, consider the relationship between number of pig's feet eaten per day in Kentucky and the amount of rainfall in Washington. Probably these two variables have nothing to do with each other, but based on the sample days you selected, you may find the following correlation: $r = -.15$. Are we to conclude that the more pig's feet eaten in Kentucky the fewer inches of rain in Washington? Probably this would not be a very valid conclusion. The observed correlation of $r = -.15$ is probably simply due to random variation in the data points we happened to observe. So how do we know when a correlation is significant (not due to chance) and when it is simply due to random error? We think you already know the answer to this one -- test the correlation for significance by comparing the calculated value to the critical value (as with every other statistical test we've done)! The only questions you should have now are what are the degrees of freedom and where is the table of critical values (Appendix A)!

But, first let's specify exactly what the null hypothesis is that we are testing. Essentially, the researcher is predicting that there is some correlation (i.e. not equal to zero) in the population between variables X and Y, so the research or alternative hypothesis can be stated as:

$$H_a: \quad \rho \neq 0$$

The parameter ρ (rho) is the value for the correlation in the population (which is what is always stated in the null and alternative hypotheses) whereas the statistic r is the correlation for a sample (in the same way as refers to the mean of a sample, but μ refers to the mean of the population in the null and alternative hypotheses). Conversely, then, we can think of the null (zero) hypothesis as stating that there is no correlation between the two variables in the population, and any correlation observed in sample data is simply due to chance:

$$H_0: \quad \rho = 0$$

Degrees of freedom for the r-test are very simple. Remember that degrees of freedom refers to how many scores are free to vary to insure a particular statistic (such as a mean) and involves the loss of one degree of freedom. You have two sets of scores (e.g., height and weight). To insure a particular mean height, the set of set of scores for height lose one degree of freedom. Similarly to insure a particular mean weight, the set of scores for weight lose one degree of freedom. With 10 pairs of scores (pairs, because 10 subjects have scores for both weight and height, so there are 10 pairs of data points), we lose one degree of freedom for the weight variable and one degree of freedom for the height variable. So, for 10 subjects we lose a total of two degrees of freedom, one for each of the two variables we are correlating, so $df = N - 2 = 10 - 2 = 8$.

Now, using the r-table (see Appendix A) we can determine the critical values for the .05 and .01 significance levels, above which the calculated value must fall in order to reject the null hypothesis (that the correlation in the population is zero). The critical values with 8 degrees of freedom at the .05 and .01 significance levels are, $r_{(8).05} = .632$ and $r_{(8).01} = .765$. Since the calculated value $r = +.60$ is not larger than the critical value $r = .632$ at the .05 significance level we do not reject H_0 and conclude that there is no relationship between the two variables (height and weight) in the population of women. This conclusion may seem what surprising given that $r = .60$ (ignoring the sign) seems like a relatively large correlation. However, the sample used in this example ($N = 10$) was very small. Remember the concept of power? To increase power (or the ability to detect significance) you can increase sample size. Note what happens to the critical values as the degrees of freedom (corresponding directly with sample size) increase -- they go down! For example, consider a sample of size 20 instead of size 10; the degrees of freedom is $df = N - 2 = 20 - 2 = 18$. The critical value at the .05 significance level is $r_{(18).05} = .444$, which is quite a bit less than .632 (the critical value with a sample size of 10). So, a larger sample size makes the likelihood that you will detect a significant correlation in the sample much higher.

Amount of Variance Explained

One of the really nice features about the correlation coefficient, r, is that it allows you to identify from one variable how much you can predict or explain about what happens with another variable. For example, if you have a correlation of -1.00 of +1.00 you know that one variable (e.g., height) perfectly explains all of the variation in the other variable (e.g., weight). That is, you can determine how much variance the two variables share with each other, in this case, 100%. If you know the person's height this perfectly explains everything you need to know about their weight. This value denoting the variance two

$0 \le r^2 \le 1$

variables share with each other is termed the **coefficient of determination** and is equal to the square of the correlation coefficient, r^2. So, if you have a correlation between two variables (e.g., parent's IQ and child's IQ) of $r = +.70$, you know that $r^2 = 49\%$ of a child's IQ can be explained by what the parent's IQ is. The other 51% of variability in a child's IQ can be explained by other factors (e.g., genetics, schooling, environment, etc.).

The coefficient of determination really provides one of the best way to visualize the concept of correlation. Consider the following two circles, each representing one variable -- amount of rainfall in Iowa per year, and number of new successful TV shows on the FOX network per year:

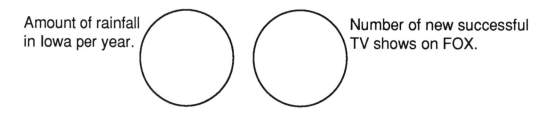

Figure 9.10 Correlation $r = 0$ and coefficient of determination $r^2 = 0$

As you can see, the two circles do not overlap at all indicating that these two variables function completely independent of one another. The correlation between these variables is $r = 0.00$ and the coefficient of determination is $r^2 = 0$, meaning the variance in one variable (rainfall) explains 0% of the variability in the second variable (successful TV shows).

Now consider the following diagram:

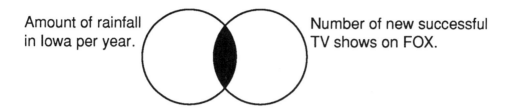

Figure 9.11 Correlation r = .30 and coefficient of determination r^2 = .09

In this case, the correlation between the two variables is r = .30 (either + or -, it doesn't matter). The coefficient of determination is r^2 = .09, meaning that 9% of the variability in successful FOX TV shows can be explained by the amount of rainfall in Iowa (or vice versa). This variance is illustrated in Figure 9.11 above. We will use these diagrams to help conceptualize correlation, since there is a direct relationship between the correlation and the variance accounted for: higher correlation (stronger magnitude of the correlation) means greater variance accounted for.

Consider another example:

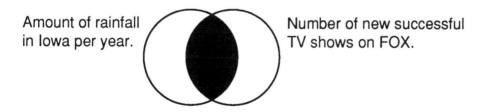

Figure 9.12 Correlation r = .80 and coefficient of determination r^2=.64

Here, the correlation is clearly higher, r = .80, with corresponding coefficient of determination of r^2 = .64. The relationship between these two variables is stronger and the percentage of variance accounted for is higher. The most extreme situation is two completely overlapping circles, with a correlation $r = +/-1.00$ and $r^2 = 1.00$, or 100% of the variability accounted for.

Correlation With Ordinal-Level Data

So far in this book we have been talking about statistical tests for interval- and ratio-level data. As you may well imagine, not all data that a researcher will ever collect will be in either of these forms. Sometimes data will be measured at the ordinal- or nominal-level. We will deal with one of the major tests (chi-square) for nominal-level data in Chapter 11, but at this point it is noteworthy to mention a special correlation procedure for data in ordinal-level form. Consider the ten women's heights and weights from earlier in this chapter:

		Height (in inches)	*Weight (in pounds)*
1.	Susan	62	108
2.	Carol	63	120
3.	Angie	64	115
4.	Mary	65	130
5.	Erin	66	137
6.	Missy	66	130
7.	Gwen	67	150
8.	Pat	68	150
9.	Lisa	68	143
10.	Ariel	70	155

Both sets of scores (for height and for weight) are in ratio-level form. But, suppose instead that we had the height rankings of the women from shortest to tallest (but not their actual height), along with their weight:

		Height Ranking	Weight (in pounds)
1.	Susan	1	108
2.	Carol	2	120
3.	Angie	3	115
4.	Mary	4	130
5.	Erin	5.5	137
6.	Missy	5.5	130
7.	Gwen	7	150
8.	Pat	8.5	150
9.	Lisa	8.5	143
10.	Ariel	10	155

When at least one of the two sets of scores is in ordinal-level form as above you cannot compute a Pearson r (because computation of this statistic assumes you can compute means and standard deviations, which you only can do with interval- or ratio-level data). In this type of situation, you should be sure that *both* sets of data are in ordinal-level form and compute a correlation statistic known as **Spearman's rho** (which is analogous to Pearson's r for interval-and ratio-level data). So, the two sets of rankings (for height and for weight) would look like the following:

	Height Ranking	Weight Ranking
Susan	1	1
Carol	2	3
Angie	3	2
Mary	4	4.5

Erin	5.5	6
Missy	5.5	4.5
Gwen	7	8.5
Pat	8.5	8.5
Lisa	8.5	7
Ariel	10	10

An Example of Correlation

Although we are not actually going to compute a correlation from scratch, you should be familiar with what it is and how to test a correlation for significance. Consider the following problem: a researcher believes that students who watch a lot of television tend to perform poorly at school. She randomly samples 52 students at a university and asks the students to indicate how many hours per day they watch television. She then goes to the Registrar's Office and obtains the grade point average (GPA) for these 52 students. She computes a correlation between these two sets of data of $r = -0.33$.

QUESTION: Is there a significant relationship between watching television and college GPA?

ANSWER:

Step 1: What are the hypotheses for this study?

The researcher is predicting that there will be some relationship (significantly different from zero) between how much college students watch television and how well they perform in school. The hypotheses for this study are:

H_a: $\rho \neq 0$

H_0: $\rho = 0$

Step 2: Compute the value of the correlation.

Since we are given the value of the correlation (which will be true throughout the book), we do not have to compute it from hand. The correlation in this problem is $r = -0.33$.

Step 3: Calculate the degrees of freedom.

Since we lose one degree of freedom for each set of scores, we can use the formula to obtain $df = N - 2 = 52 - 2 = 50$ degrees of freedom.

Step 4: Look up the critical value at the $p < .05$ level.

Using the r-table, we find the critical value for 50 degrees of freedom at the .05 significance level is:

$r_{(50).05} = .273$.

Step 5: Compare the calculated value to the critical value.

Since the calculated value $r = -0.33$ is larger (in absolute terms) than the critical value $r = .273$, we can reject H_0 at the $p < .05$ significance level and conclude that there is a significant relationship between television viewing and college GPA, with a Type I error of .05 (meaning there is a 5% chance that there actually is no relationship between these two variables in the population of university students, and the correlation we observed in the sample of 52 students occurred purely by chance).

Step 6: Look up the critical value at the $p < .01$ level.

Using the r-table, we find the critical value for 50 degrees of freedom at the .01 significance level is:

$r_{(50).01} = .354$.

Step 7: Compare the calculated value to the critical value.

Since the calculated value $r = -0.33$ is not larger (in absolute terms) than the critical value $r = .354$ at the .01 level, we cannot reject H_0 at the $p < .01$ significance level. We must end up rejecting H_0 at the $p < .05$ level and living with a Type I error of .05.

Step 8: State the final conclusion both statistically and in words.

Statistically, the decision is to reject H_0 at the $p < .05$ level. We conclude that there is a significant (not due to chance) relationship between amount of television watched and college GPA of university students, with a 5% chance that there really is no relationship at all.

Partial Correlation and Multiple Correlation

Not all variables are neatly correlated in a simply one-to-one relationship. Most variables are related to multiple other variables simultaneously. Consider the following relationship between three variables related to college life (amount of time spent studying, amount of time spent partying, and college grade point average):

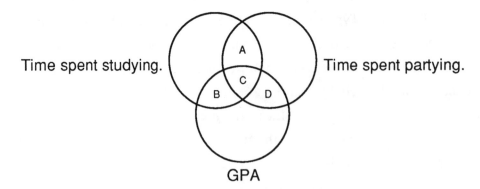

Figure 9.13 A Venn diagram graphically illustrating partial and multiple correlation

The overlapping circles represent the interrelationships of these three variables with each other. For example, consider the simple correlation between the two variables amount of time spent studying and amount of time spent partying. Assume this correlation is $r = -0.45$ (since presumably the more time you spend studying the less time you have available for partying, and vice versa). This correlation is represented (in terms of its coefficient of determination, $r^2 = 0.20$) as the areas indicated by the letters A and C in Figure 9.13 above, so $r^2 = A + C$. Now consider the representation for the correlation of time spent studying with GPA, $r = +.38$, area $B + C$ in the Figure 9.13. We can represent each of the simple correlations (actually, the variances r^2) in the diagram above as follows:

$$r^2_{(study)(party)} = A + C$$
$$r^2_{(study)(GPA)} = B + C$$
$$r^2_{(party)(GPA)} = C + D$$

Notice in each of these simple correlations that there is some overlap. For example, the correlation between time spent studying and GPA is represented by the area $B + C$ and the correlation between time spent partying and GPA is represented by the area $C + D$. These two simple correlations share the area C; that is, area C represents the correlation of all three variables with each other. But what if we want to examine the pure relationship between time spent studying and GPA? In other words, what is the relationship between time spent studying and GPA removing the contribution that any other variable (e.g., time spent partying) has on this relationship? Well, if we remove the area C from the simple correlation represented by $r^2_{(study)(GPA)} = B + C$, we end up with area B. No other variables impede on area B. Area B is the area shared exclusively between the variable time spent studying and the variable GPA. So, if we <u>remove or partial out the effect of any third variables</u>, we end up with the **partial correlation** between two variables, which is the exclusive correlation between these two variables. So, area B represents the amount of variance in GPA that is explained by time spent studying only. In contrast, area C represents the amount of variance in GPA that is accounted for by the joint relationship between time spent studying and time spent partying. So, we can represent the partial correlation between time spent studying and GPA partialling out (removing) the effects (shared correlation) of time spent partying as $r^2_{(study)(GPA) \cdot party} = B$. Similarly, we can represent all of the partial correlations of each of the variables with each other, partialling out the effects of the third variable as the following:

$$r^2_{(study)(party) \cdot GPA} = A$$
$$r^2_{(study)(GPA) \cdot party} = B$$
$$r^2_{(party)(GPA) \cdot study} = D$$

Notice that with the partial correlation the symbol • means "partialling out (removing) the effects of." So, $r^2_{XY \cdot Z}$ reads "the partial correlation of variable X with variable Y partialling out the effects of variable Z."

Finally, what if we want to know about the relationship of more than two variables, that is the **multiple correlation**? For example, what is the relationship (multiple correlation) of GPA with *both* time spent studying and GPA. What do both variables, both independently and together, contribute to GPA? Well, we know that time spent studying contributes area B + C (in terms of variance) and time spent partying contributes area C + D. Since C is not two unique areas we should only count it once. Thus, we can represent the multiple correlation of GPA with time spent studying and time spent partying as $r^2_{GPA \cdot (study)(party)} = B+C+D$. Similarly, we can look at the multiple correlation of each variable with the two other variables:

$$R^2_{\text{study}\cdot(\text{study})(\text{party})} = B + C + D$$
$$R^2_{\text{GPA}\cdot(\text{party})(\text{GPA})} = A + C + B$$
$$R^2_{\text{party}\cdot(\text{study})(\text{GPA})} = A + C + D$$

Notice that with the multiple correlation the symbol • means "with". So, $R^2_{X \cdot YZ}$ reads as "the multiple correlation of variable X with both variables Y and Z". Notice too, the use of a capital R. The small letter r is used for correlation between two variables (simple correlation and partial correlation) and the big letter R is used for correlation between more than two variables (multiple correlation).

Chapter 9 Problems

1. If you want to examine the relationship or association between two variables which statistical test should you conduct?

2. The following questions refer to characteristics of a simple correlation:
   a. What is the range of values that a correlation coefficient can possibly assume?
   b. What does the sign of a correlation coefficient tell you?
   c. What can be inferred about the relationship between two variables when there is a positive correlation: as one variable increases, the other variable (*increases* or *decreases*)?
   d. What can be inferred about the relationship between two variables when there is a negative correlation: as one variable increases, the other variable (*increases* of *decreases*)?
   e. What does the magnitude of a correlation coefficient reveal?

3. For each of the following correlations depicted below, indicate the approximate magnitude of the correlation (*none, low, moderate*, high, or *perfect*) and the direction of the correlation (*positive, negative*, or *none*)?

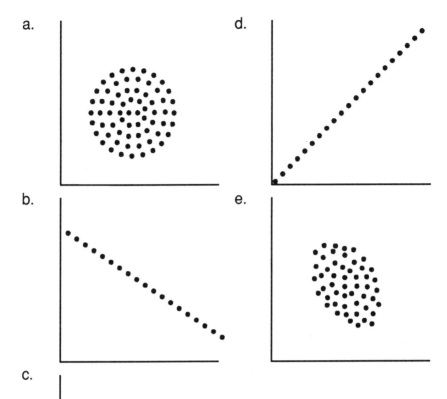

4. What is a simple correlation and how is it symbolized?

5. If at least one of the variables is in ordinal-level form, the appropriate computation of the correlation is the (*Pearson r* or *Spearman's rho*).

248 CORRELATION

6. For each of the following pairs, indicate which is the *stronger* correlation:

   a. $r = +0.56$ or $r = +0.78$
   b. $r = -0.13$ or $r = -0.02$
   c. $r = +0.86$ or $r = -1.00$
   d. $r = -0.92$ or $r = +0.42$
   e. $r = -0.21$ or $r = 0.00$
   f. $r = -1.00$ or $r = +1.00$

7. A researcher wishes to assess the relationship between computer consultants' deteriorating vision and the amount of time they spend in front of a computer screen. Thirty consultants' weekly amount of time spent typing on a computer and vision are measured. The data revealed that the correlation between these two variables is $r = -0.43$.

   a. State the null and research hypotheses.

   b. Conduct a test of the correlation coefficient.

   c. What is the magnitude and the direction of this correlation?

   d. State your decision in words.

8. A researcher has two assistants observe 52 subjects engaged in a five-minute conversation. One observer rates each subject's degree of talkativeness and the other observer rates each subject's intelligence. The correlation between degree of talkativenss and perceived intelligence is $r = +0.24$.

   a. State the null and research hypotheses.

   b. Conduct a test of the correlation coefficient.

c. What is the magnitude and the direction of this correlation?

d. State your decision in words.

9. A researcher suspects a relationship between number of days of rainfall per year and the number of absentee days from work. Ten people are randomly selected from cities across the country and the amount of rainfall in each person's city for 1991 is correlated with the number of days each person was absent from work. A correlation of r = +0.78 is obtained.

a. State the null and research hypotheses.

b. Conduct a test of the correlation coefficient.

c. What is the magnitude and the direction of this correlation?

d. State your decision in words.

10. What statistic indicates the percentage of variance two variables have in common? How is this symbolized?

11. Find r^2 for each correlation below and explain what the shaded section of each diagram represents.

250 CORRELATION

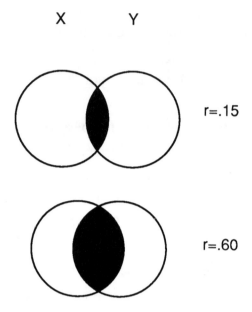

12. For each of the following, indicate the magnitude (*none, low, moderate, high,* or *perfect*) and direction (*positive, negative* or *none*) of the correlation; and the percentage of variance accounted for:

   a. +0.84
   b. -0.13
   c. -1.00
   d. +0.22
   e. 0.00
   f. +0.92
   g. -0.57

13. What statistical test examines the association among at least three variables?

14. What is the symbol for the multiple correlation of variable X with variables Y and Z? What is the symbol for the coefficient of determination for the multiple correlation? What does this coefficient tell you?

15. What statistical test examines the association of two variables, removing the association of all other variables?

16. What is the symbol for the partial correlation of variables U and V partialling out the effect for variable Y? How would you symbolically represent the coefficient of determination for the partial correlation? What does this coefficient tell you?

17. Refer to the following diagram:

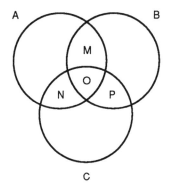

a. What area represents r^2_{AC}?
b. What area represents $R^2_{A \cdot CB}$?
c. What area represents $r^2_{AC \cdot B}$?
d. What area represents $R^2_{B \cdot AC}$?
e. What area represents r^2_{BC}?
f. What area represents $r^2_{BC \cdot A}$?
g. What area represents $R^2_{C \cdot AB}$?
h. What area represents $r^2_{AB \cdot C}$?
i. What area represents r^2_{AB}?

CHAPTER 10
REGRESSION

From Correlation to Regression

Whether you know it or not, we already have introduced the concept of regression in the previous chapter on correlation. In fact, this is probably one of the easiest (and shortest!) chapters in the book (good -- time left for pizza!). Recall that simple correlation is a measure of how two variables are related to one another, both in terms of direction (positive or negative) and magnitude (strength, ranging from 0.00 to +/-1.00). For example, a moderate positive correlation looked like the following:

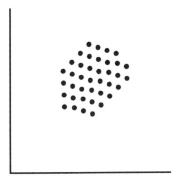

Figure 10.1 A moderate positive correlation

Notice that the data points (each representing a pair of scores, such as height and weight) are generally in an upward-sloping direction -- the taller someone is the more they tend to weigh. Recall from the last chapter that we drew a line through these data points that seemed to best fit the data points (that is, by minimizing, on average, how far each data point is away from the line). This line is called the **line of best fit** (so named for obvious reasons) or the **regression line**.

The regression line (and any line for that matter) is characterized by two things: a slope and a constant. The slope of the line simply refers to how much the line is slanted upward (with a positive slope) or downward (with a negative slope). The slope is referred to with the statistic, b, or beta value. The following diagram is an example of a group of possible data points for height and weight, with the correlation, r, and the corresponding beta value representing the slope of the line:

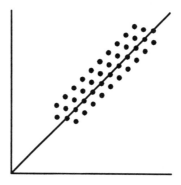

Figure 10.2 High, positive correlation $r = .80$ and line of best fit (with slope $b = .50$)

You should note that the slope of the line is not necessarily directly related to the correlation between two variables. For example, both of the following graphs illustrate perfect positive correlations (since all of the data points fall exactly on the line), with $r = +1.00$:

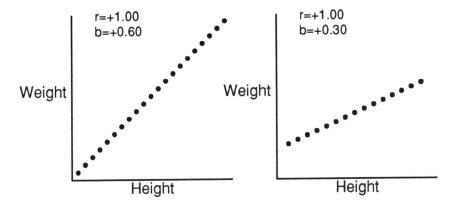

Figure 10.3 Perfect positive correlations with steeper slope, $b = .60$ (left) and flatter slope, $b = .30$ (right).

However, the graph on the left has a line with a much steeper slope ($b = +0.6$) than does the line in the graph on the right ($b = +0.3$).

The second characteristic of a regression line is the constant. The constant, a, is simply the place where the line crosses the y-axis. So, for example, consider the following graph of the two variables hours of exercise per day and hours spent sleeping per day, with a positive correlation of $r = +0.20$ and a slope of the regression line of $b = +0.4$:

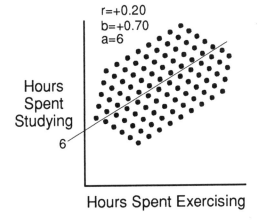

Figure 10.4 Low, positive correlation ($r = .20$) with slope, $b = .4$, and constant, $a = 6$

Notice that the line crosses the y-axis at the point of 6 hours of sleep per day. On average, a person who exercises zero hours per day sleeps 6 hours per day. Although we won't actually compute the slope or constant by hand, you should know that they are a function of the correlation between the two variables, the means of each of the two variables (e.g., hours of sleep, hours of exercise), and the standard deviations of the two variables.

Using the slope of the regression line, b, and the constant, a, (where the line crosses or intercepts the y-axis), we can produce an equation that represents this line:

$$Y = bX + a$$

The value for the second variable, Y, equals the slope of the line b, multiplied by the value for the first variable, X, plus the constant, a. This equation allows us to *predict* one variable, Y, from another variable, X, a concept known as **linear regression** (called linear because this is the equation for a straight line). Consider the example above with exercise and sleep (see Figure 10.4). Given the slope ($b = +.4$) and constant ($a = +6$) for this line, we obtain the following equation for the regression line:

$$Y = +0.4X + 6$$

We already know that someone who exercises zero hours per day sleeps 6 hours per day (according to the regression line computed as the line of best fit for some set of sample data). We can see this by plugging in the value of zero into the regression equation:

$$Y = +0.4X + 6 = +0.4(0) + 6 = 0 + 6 = 6$$

Now suppose that we know that someone exercises one hour per day. What is the best prediction, based on the regression line, for how many hours of sleep this person gets per day? Using the regression equation:

$$Y = +0.4X + 6 = +0.4(1) + 6 = 0.4 + 6 = 6.4$$

We can see these predictions in the following graph:

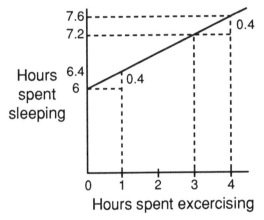

Figure 10.5 Prediction of hours spent sleeping (Y-variable) from hours spent exercising (X-variable), with the equation for the regression line, $Y = +.4X + 6$

Notice that by increasing one unit of the X-variable (increasing one hour of exercise) the value of the Y-variable (hours of sleep) went up by the slope of the line (+0.4). Similarly, note what happens when there is an increase from 3 to 4 hours of exercise: the number of hours spent sleeping increases from 7.2 to 7.6 -- the slope of the line! Had this been a downward sloping line, an increase of one in the X-variable would have led to a decrease of the value of the slope of the line in the Y-variable -- certainly a neat characteristic of lines!

Assessing the Regression Line for Significance

With regression we are predicting one variable from another (i.e. predicting sleep from exercise). But, how do we know how accurate this prediction is, or if it is really no better than chance guessing? Consider these two extreme cases:

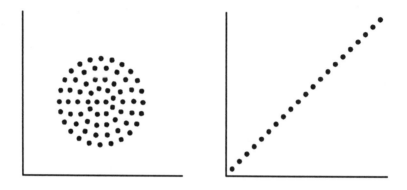

Figure 10.6 Perfect correlation, r = +1.00 (or r = -1.00) means perfect prediction (right), whereas no correlation r = 0.00 means no prediction (left)

In the graph on the left, the correlation between the two variables (hours of exercise and hours of sleep per day) is $r = +1.00$, a perfect correlation (notice that all of the data points fall exactly on the line!). If we use the regression equation to predict someone's sleep from their exercise value, we will predict this value with 100% accuracy! Why? Because the relationship between the two variables is perfect. Recall from the last chapter that the coefficient of determination for a perfect correlation is $r^2 = 1.00$, meaning that 100% of the variance in one variable is accounted for by the other variable. If we know the value of one variable (i.e. hours of exercise) we know everything (100%) about the variability in the other variable (i.e. hours of sleep), so we have perfect prediction.

Now consider the other extreme case on the right-hand side of Figure 10.6. This is a situation in which the correlation between the two variables is non-existent, $r = 0.00$. There is no relationship whatsoever between the two variables. If we use the regression equation to predict someone's sleep from their exercise value in this situation, we are predicting no better than chance -- essentially we're guessing. Knowing how much someone exercises per day tells us absolutely nothing about how much they sleep per day. We can see this by noting that the coefficient of determination is $r^2 = 0.00$, meaning that 0% of the variance or variability in one variable is explained by the other variable.

So, what about correlations between these two extremes of perfect prediction and chance guessing. How do you know when you are predicting *significantly* better than chance? Well, you already know the answer to this one. Whenever the correlation between the two variables is significant, the regression equation also will be significant, meaning that using the regression line to make predictions *is* better than chance guessing with a significant correlation between the two variables.

Prediction From Regression and Error

Notice that in all situations except a perfect correlation ($r = +1.00$ or $r = -1.00$) not all of the data points (pairs of scores) fall exactly on the line of best fit; that is, there is some error. Granted, you try to minimize this error by drawing the regression line such that it minimizes, on average, how far away each data point is from the line; however, there is still some error. When you make a prediction using the regression line, with its built-in error, you probably will be off a little in your prediction (unless the correlation is perfect). That is, the value you predict may not be exactly the actual value of the variable. For example, if we use a set of sample height and weight data with a correlation of $r = +0.75$ and use the resulting regression line for prediction, we may predict that someone who is 5'9" tall weighs 160 pounds. However, due to the fact that the relationship between height and weight is not perfect (the correlation is not +1.00 or -1.00), the person actually may weigh 163 pounds. This difference between the weight we predicted (160 pounds) and the actual weight of the person (163 pounds) is error. When we look at these differences (error) across many data points, an error value can be computed based on the correlation of the two variables which is referred to as the **standard error of the estimate (SEE)**. Correlation and error are inversely related to each other, meaning that the higher the correlation (the closer to perfect prediction) the lower the error and vice versa. This makes sense since a higher magnitude for a correlation means that the data points are closer to the regression line, so the actual and predicted values are also closer. The standard error of the estimate is useful as a measure of how accurate the prediction is. While the significance of the correlation tells whether we are predicting better than chance guessing, it does not give us a sense for exactly how good this prediction is, which the standard error of the estimate does. Higher correlation generally means a lower standard error of the estimate and a more accurate prediction.

An Example of Regression

Let's now try a brief example of linear regression. Suppose an educational researcher is interested in predicting college GPA from SAT scores (a common assumption for admissions is that a student's SAT scores are a predictor of college performance). He samples 62 college students in an educational measurement class he is teaching and obtains both their college GPA and their SAT scores. He computes the correlation between these two variables to be $r = +0.28$. Using this correlation and the means and standard deviations of both sets of scores (GPA and SAT) he calculates the following regression equation: GPA = (+0.0023) SAT + 0.3.

QUESTION: Based on this regression equation, what is the GPA prediction for a student who had a 700 SAT score? What about someone with a 1500 SAT score?

ANSWER: Just plugging in the predictor (SAT) values, we obtain the following predictions for college GPA:

GPA = (+0.0023)SAT + 0.3 = (+.0023)(700) + 0.3 = 1.61 + 0.3 = 1.91

GPA = (+0.0023)SAT + 0.3 = (+.0023)(1500) + 0.3 = 3.45 + 0.3 = 3.75

So, if a student has an SAT score of 700, the researcher's best prediction is that s/he will obtain a college GPA of 1.91, and if the student's SAT score is 1,500 s/he is predicted to obtain a college GPA of 3.75.

QUESTION: Is the researcher predicting any better than chance guessing (i.e. is the regression equation significant)?

ANSWER: The answer to this question simply involves assessing whether the correlation is significant. The correlation, $r = +0.28$ with 62 subjects, has $df = N - 2 = 62 - 2 = 60$ degrees of freedom. At the .05 significance level with 60 degrees of freedom, the critical value is $r_{(60).05}$ is $r = .250$ and at the .01 significance level the critical value is $r_{(60).01} = .325$ (from the r-table, see Appendix A). Since the calculated value $r = 0.28$ is larger than the critical value $r = 0.25$ at the .05 significance level we can reject H_0 (that there is no relationship between the two variables) and conclude that there a significant relationship between SAT score and college GPA with a 5% chance (Type I error) that there is actually no correlation between these variables at all. Testing the correlation for significance at the .01 level, we find that the calculated value $r = 0.28$ is not larger than the critical value $r_{(60).01} = .325$, so we end up rejecting

H_0 at the $p < .05$ significance level with a 5% chance of being wrong. So, the regression equation predicts significantly better than chance, with a 5% probability that the equation actually predicts no better than chance guessing.

CHAPTER 10 PROBLEMS

1. What is the purpose of linear regression?

2. What is the name of the line drawn through pairs of data points on a graph that represent the correlation between two variables?

3. What is the equation for a regression line for linear regression?

4. What is the symbol for the slope of a line and what does it mean?

5. What is the symbol for the constant of a line and what does it mean?

6. For each of the following regression equations and X-values, give the best prediction of the Y-value:
   a.   Y = +0.84X + 9   X = 3
   b.   Y = 32X + 5,   X = 8
   c.   Y = 10X + 0.2,   X = 5.3
   d.   Y = 0.33X + 0.55,   X = 1

7. What information does the standard error of the estimate (SEE) provide?

8. Is the relationship between the correlation coefficient, r, and the standard error of the estimate, SEE, positive or negative?

9. If the correlation between X and Y is high, will the SEE be small or large?

10. For each of the following pairs of correlations, indicate which one would have the lower standard error of the estimate (i.e. would yield the most accurate prediction):

   a. r = +0.34,   or   r = -0.21
   b. r = 0.00,   or   r = -1.00
   c. r = +0.56,   or   r = +0.67
   d. r = -0.49,   or   r = + 0.32
   e. r = -0.98,   or   r = -0.56

11. Indicate for each of the following instances of linear regression whether the regression equation is significant (that is, whether it predicts significantly better than chance):

   a. A researcher samples 30 individuals and finds the correlation between the number of cigarettes smoked per day and how many hours per day the person sleeps is r = -0.28. Can the researcher predict better than chance how much a person will sleep per day based on knowing how many cigarettes s/he smokes?

   b. Twenty-two subjects participated in a study of how much time workers spend chatting with their colleagues during the day and how many cups of coffee they drink. The correlation was r = + 0.47. Could a manager predict better than chance how much time an employee will spend gossiping with coworkers based on knowing how many cups of coffee s/he drinks per day?

CHAPTER 11
CHI-SQUARE ANALYSIS

Parametric vs. Nonparametric Tests

Up until now we have been dealing with statistical tests that can be used to analyze interval- or ratio-level data (with the exception of Spearman's rho as the correlational procedure for ordinal-level data). The z-test, t-test, F-test, Pearson r, and linear regression are all statistical techniques for analyzing interval- or ratio-level data. These data analysis techniques as a group can be referred to as **parametric statistics** because they deal with characteristics of populations (such as means and standard deviations), which are referred to as parameters. [handwritten: very strict assumptions concerning distributions] However, whenever you have data that are measured at the nominal- or ordinal-level you cannot estimate parameters for populations. Thus, statistical techniques that do not deal with estimates of populations and involve nominal- or ordinal-level data are referred to as **nonparametric statistics**. In this chapter we will be dealing with one of the major nonparametric statistical tests for nominal-level data: the chi-square test.

The Chi-Square (χ^2) Test

The chi-square test focuses on nominal-level data. Instead of asking, like the t- and F-tests did, if there is a significant difference in means for groups the χ^2-test asks if there is a significant difference in the number of people (or frequency of observations) per category. Basically, the concept is similar to the t- and F-tests, except instead of working with means of groups you are working with the total number of subjects (or observations) per group. Let's start with a simple example. Suppose your professor is interested in students' preference for old television programs. Do more students prefer *I Dream of Jeannie* or *Bewitched*? The professor asks the 50 students in her introductory statistics class to write down on a piece of paper which old TV show they prefer. After she collects the data she counts the number of students who prefer each show and obtains the following data:

I Dream of Jeannie	*Bewitched*
35	15

What are the hypotheses for this study? Well, the researcher is actually asking a research question in this case: is there a difference in the frequency of occurrence in the two categories (do students differentially prefer one or the other program)? However, we are going to state this as a research hypothesis and null hypothesis. Essentially, the researcher's hypothesis would be that there is a significant difference between the number of students who prefer *I Dream of Jeannie* and the number of students who prefer *Bewitched*, with the corresponding null hypothesis that there is not a significant difference in the number of students preferring the two programs. The null and alternative hypotheses for the chi-square test involve frequencies of observations expected by chance, but are somewhat easier to understand stated in words rather than statistically, so we will treat them in this way.

As with the t- and F-test, the question ultimately becomes is the difference we observed a significant difference? Clearly the numbers 35 and 15 are not the same. But, is this difference due purely to random factors (the 50 people who happened to show up on the day the professor queried students, the students' mood, etc.), or is it actually the case that statistics students prefer *I Dream of Jeannie* significantly more than *Bewitched*? You should notice a distinct similarity between the type of questions we are asking here and the type of questions we asked with the t-test and ANOVA. The major difference is that with the latter we were working with means (interval- or ratio-level data) and with the current problem we are working with nominal-level data (frequency of occurrence of observations in categories).

The example we have illustrated above is called a **one-sample chi-square** problem, or difference among categories, because we are looking at differences in people across categories for only one sample (introductory statistics students in this professor's class). Suppose that instead the researcher was interested in whether males or females in her class differed in their preference for *I Dream of Jeannie* or *Bewitched*. We might obtain data for the 50 students that looked the following:

	I Dream of Jeannie	*Bewitched*
Males	20	5
Females	15	10

Now we might ask if is there a difference in males' and females' preference for these two old TV programs. This question is now assessing a difference among samples (males and females) and is referred to as a **multiple-sample chi-square** problem. When we have two variables (e.g., gender, type of TV show) for a multiple-sample chi-square problem presented like above we refer to this as a **contingency table**. We identify contingency tables according to the number of levels for each of the variables in exactly the same way as we did with factorial designs. This problem is an example of a 2 x 2 contingency table: there are two variables so there are two numbers. The first number (representing gender) has 2 levels (male, female), and the second number (representing TV show) has 2 levels (*I Dream of Jeannie*, *Bewitched*).

Calculating the χ^2 Value

The χ^2-test is conducted by comparing the frequencies that you actually observed with what you would expect by chance (theoretically) if there were truly no difference in frequencies. For example, let's consider the simple one-sample chi-square problem with the 50 students' preference for one of the two old TV programs:

<u>I Dream of Jeannie</u>   <u>Bewitched</u>
35   15

These are the **observed frequencies**: 35 of the 50 students indicated that they prefer *I Dream of Jeannie* and the other 15 of the 50 students indicated that they prefer *Bewitched*.

If there was no difference in students' preference for these two programs, how many students (out of these same 50 students) would you expect to prefer each program? The answer is 25 each:

<u>I Dream of Jeannie</u>   <u>Bewitched</u>
25   25

These are the **expected** or **theoretical frequencies**: if there was no difference in students' preference for the two TV shows 25 would prefer *I Dream of Jeannie* and 25 would prefer *Bewitched*. The theoretical or expected frequencies for the one-sample case are equal to the sum of the observed frequencies across all categories (e.g., 50) divided by the total number of categories either rows or columns, depending on how you lay your data out (e.g., 2 columns in our example). Essentially, each category would have the same number of expected observations if there were truly no difference in the categories. The more the observed values deviate from the expected or theoretical values the more likely it is that there is a significant (not due to chance) difference in students' preference for the two programs. As with means note that the observed values may deviate from the theoretical values somewhat (say 26 students prefer *I Dream of Jeannie* and 24 students prefer *Bewitched*), but there would not be a *significant* difference in students' preference for these two programs. The difference between the observed and expected values is simply due to chance or random factors. However, the more the observed values deviate from the expected values, the more likely it is that there is a significant difference in the frequencies of observations in categories.

The χ^2 statistic is a measure of how much the observed frequencies deviate from the expected frequencies. The more the observed values deviate from the theoretical values, the higher the χ^2-statistic and the more likely that there is a significant difference among categories. This deviation between observed values (O) and theoretical values (T) is computed using the following formula for the χ^2-statistic:

$$\chi^2 = \Sigma \frac{(O-T)^2}{T}$$

Notice in the numerator of the equation that we are taking the difference between the observed value (O) and the theoretical value (T); that is, creating a difference or deviation score. A larger deviation in these two values means a higher χ^2-value. Note, two, that we are squaring this difference -- $(O-T)^2$. If you recall the computation for sum of squares for the ANOVA we took the difference between the group mean and the grand mean (for SS_B) and squared these values for between individual scores and the group mean for SS_W. This was done in order to get rid of any negative values, so that when you sum these differences the positive and negative values do not cancel each other out. Exactly the same principle applies to the χ^2, so the differences between observed and theoretical values are squared to get rid of negative numbers. This squared difference is then divided by the theoretical value to get something akin to a mean (sort of like dividing sum of squares by degrees of freedom). The Σ symbol (which you may recall from the equation for a mean) means "sum up". So, computing the χ^2-value involves summing up values for each of the categories, comparing the observed with the theoretical values.

Testing the One Sample χ^2 for Significance

Let's compute the χ^2-statistic for the one-sample TV program scenario. Recall the observed and theoretical frequencies:

	I Dream of Jeannie	*Bewitched*
Observed Values	35	15
Theoretical Values	25	25

Now using the χ^2-formula, let's compute the χ^2-statistic for these data:

$$\chi^2 = \Sigma \frac{(O-T)^2}{T} = \frac{(35-25)^2}{25} + \frac{(15-25)^2}{25} =$$

$$= \frac{10^2}{25} + \frac{-10^2}{25} = \frac{100}{25} + \frac{100}{25} = 4+4 = 8$$

So, the calculated value of χ^2 is 8. Now what? Well, just as with all of the other statistical tests we have done, once you have computed the calculated value you must compare it to some critical value.

Recall that a higher χ^2-value means that there is a larger difference between the observed and theoretical values, so you are more likely to detect a significant difference with a higher χ^2-value (which was true for all of the other statistical tests we have done as well). This is analogous to, for example, the t-test. The further apart the two sample means (the larger the difference in means), the higher was the test statistic t, and the more likely you were to be able to detect a significant difference and reject H_0. So, we need to obtain a calculated χ^2-value that is larger than some critical χ^2-value at the $p < .05$ or $p <$

.01 significance level in order to reject the null hypothesis. These critical values can be located in the χ^2-table in Appendix A. By now you should not be surprised to see this table organized according to degrees of freedom. As with most statistical tests (except the theoretical z-test), the critical value is a function of the number of groups and/or number of subjects you have. For the chi-square test, the critical value is a function of the number of groups (categories) you have. In the one-sample case the degrees of freedom is equal to the total number of categories you have minus one; this makes sense that all but the last group is free to vary. The last one is predetermined, just like with group means for SS_B for the F-test. Depending on how you organized your table (in rows, R, or columns, C), the degrees of freedom formula for the one-sample case would be:

$$df = (R - 1)$$
or $$df = (C - 1)$$

For our example, we have two categories organized in two columns, so $df = (C - 1) = (2 - 1) = 1$. There is one degree of freedom for this problem.

We can use the χ^2-table in Appendix A to determine the critical values for a χ^2 with 1 degree of freedom at the $p < .05$ and $p < .01$ levels:

$$\chi^2_{(1).05} = 3.84$$
$$\chi^2_{(1).01} = 6.63$$

Now, as with all statistical tests, let's compare the calculated value with the critical value, first at the .05 level, and then, if we can reject H_0, at the .01 level. If the calculated value exceeds the critical value then we can reject H_0. Since the calculated value $\chi^2 = 8.00$ is larger than the critical value $\chi^2_{(1).05} = 3.84$, at the .05 significance level we can reject H_0 at the $p < .05$ level and conclude that there is significant difference in students' preference for *I Dream of Jeannie* and

Bewitched with a Type I error (chance of being wrong) of .05. Can we do any better than a 5% error rate? Since the calculated value $\chi^2 = 8.00$ is larger than the critical value $\chi^2_{(1),.01} = 6.63$ at the .01 significance level we can reject H_0 at the $p < .01$ level and conclude that there is a significant difference in students' preference for the two TV shows with only a 1% chance of being wrong (that there actually is no difference, in students' preference for the two shows and the observed values are purely due to chance).

Testing the Multiple-Sample χ^2 for Significance

Conducting a chi-square test with a multiple-sample problem is really no different than for a one-sample problem. Let's look at the multiple-sample version of the TV show problem. We have the same 50 students, except now they are divided into male and female samples with the following data:

	I Dream of Jeannie	*Bewitched*
Males	20	5
Females	15	10

Again, we need to look at what we would theoretically expect to happen, in this case if there were no difference between the popularity of these shows for the two samples. That is, do males or females differentially prefer one of the two shows? Is *I Dream of Jeannie* more popular among males than among females? Again, the null and alternative hypotheses are easier to understand in words (the statistical formulation of these hypotheses involves the frequencies expected due to chance). We can think of the research hypothesis as predicting that there is a difference in males' and females' (the samples) preference for *I Dream of Jeannie* and *Bewitched* (the categories). Conversely, the null hypothesis is that there is no difference in males' and females' preference for the two TV shows. Note that with the hypotheses for the one-sample chi-square case we were predicting a difference in the number of observations across the different categories. In contrast, in the multiple-sample chi-square we are predicting a difference in the number of observations across the different categories (TV shows) *for the different samples* (males and females).

To compute the expected or theoretical values we need to maintain the total number of students that we actually observed. That is, there still need to be 50 students overall in the study. We observed 25 males and 25 females participating, so we still need this to be true. We need to now consider the total number of people who preferred each show -- 35 people preferred *I Dream of Jeannie* and 15 people preferred *Bewitched*. We still need those total values to be true when computing the theoretical (expected) values. We must maintain the exact same frequencies on the margins (sum of rows, sum of columns) with the theoretical table as the observed table. The observed table looks like the following:

	I Dream of Jeannie	*Bewitched*	
Males	20	5	25
Females	15	10	25
	35	15	50

Now in creating the theoretical table we must end up with the same observed marginal frequencies:

	I Dream of Jeannie	*Bewitched*	
Males			25
Females			25
	35	15	50

Now all we need to determine is what the specific cell frequencies should be for the theoretical table. Well, looking at the total number of males and females (25 and 25), if we were to expect no difference in their preference for the two TV shows we would expect the same number of males as females to prefer *I Dream of Jeannie* and the same number of males as females to prefer *Bewitched*. With 35 total people preferring *I Dream of Jeannie*, and no differences between the two samples, half should be males and half should be females; (35/2 = 17.5). Similarly, with 15 total people preferring *Bewitched*, half should be males and half should be females (15/2 = 7.5). So, the theoretical table would look like the following:

	I Dream of Jeannie	*Bewitched*	
Males	17.5	7.5	25
Females	17.5	7.5	25
	35	15	50

If we sum the number of males we should get 25: 17.5 + 7.5 = 25, and similarly 25 for females. If we sum the number of males and females who prefer *I Dream of Jeannie* we should get 35: 17.5 + 17.5 = 35, and, similarly 15 for *Bewitched*. This was a pretty easy example of computing the theoretical frequencies because we had the same number of people in each sample (25 males, 25 females). But, not all samples will be of equal size. So, a general formula for computing the theoretical or expected frequency for the multiple-sample case is:

$$T = \frac{R \times C}{G}$$

$$f_e = \frac{\text{row total} \times \text{column total}}{\text{grand total}}$$

where T is the theoretical frequency for a particular cell, R is the marginal frequency for the row that the cell is in (or total observed frequency for the row), C is the marginal frequency for the column that the cell is in (or total observed frequency for the column), and G is the grand total number of observations in the entire study. So, let's use this formula to compute what the theoretical frequency should be for males' preference for *I Dream of Jeannie* (we already know from above without the formula that the answer is 17.5):

$$T = \frac{R \times C}{G} = \frac{25 \times 35}{50} = \frac{875}{50} = 17.5$$

Similarly, let's compute the theoretical frequency for males' preference for *Bewitched* using this formula (again, we know from above that the answer should be 7.5):

$$T = \frac{R \times C}{G} = \frac{25 \times 15}{50} = \frac{375}{50} = 7.5$$

So, the formula does work. You will probably find it easier to compute the theoretical frequencies for the multiple-sample case using the formula when the samples (e.g., males and females) are of unequal size. If the samples are of equal size, then you can divide the total number of observations in the category (e.g., *I Dream of Jeannie*, *Bewitched*) by the total number of samples (two in the example), to come up with an even distribution across samples. But, of course, the formula works with equal sample sizes as well (and may be a good check that you divided frequencies correctly among categories).

Now that we have both the observed and the theoretical values, let's compute the χ^2-value. Using the χ^2-formula, we compare the observed to the theoretical value for each of the cells in the contingency table (four in our example):

CHI-SQUARE ANALYSIS

$$\chi^2 = \Sigma \frac{(O-T)^2}{T} = \frac{(20-17.5)^2}{17.5} + \frac{(15-17.5)^2}{17.5} + \frac{(5-7.5)^2}{7.5} + \frac{(10-7.5)^2}{7.5} =$$

$$= \frac{2.5^2}{17.5} + \frac{-2.5^2}{17.5} + \frac{-2.5^2}{7.5} + \frac{2.5^2}{7.5} = \frac{6.25}{17.5} + \frac{6.25}{17.5} + \frac{6.25}{7.5} + \frac{6.25}{7.5}$$

$$= 0.36 + 0.36 + 0.83 + 0.83 = 2.38$$

So, the computed χ^2-statistic is $\chi^2 = 2.38$.

Now we need to determine if this value is large enough to be able to reject H_0 and conclude that there is a significant difference in the preference for the two TV shows among males and females. The only thing we need yet is to determine the number of degrees of freedom for this problem -- this is fairly easily done. Essentially we have two variables here: gender and type of TV program. We lose one degree of freedom from each variable. For gender, there are two levels, male and female, minus one degree of freedom lost equals one. For type of TV program, there are also two levels, *I Dream of Jeannie* and *Bewitched*, minus one degree of freedom lost equals one. Then we simply multiply the two degrees of freedom for the two variables together: $df = 1 \times 1 = 1$. We can state this idea in a formula, where one variable is designated in rows (in the example, the sample variable, gender), and the second variable is designated in columns (in the example, the categorical variable TV show):

$$\boxed{df = (R - 1) \times (C - 1)} = (2 - 1) \times (2 - 1) = 1$$

So, we are conducting a χ^2 with 1 degree of freedom. Using the χ^2-table, we obtain the following critical values:

$\chi^2_{(1).05} = 3.84$

$\chi^2_{(1).01} = 6.63$

Comparing the calculated value to the critical value, we find that the calculated value $\chi^2 = 2.38$ is not larger than the critical value $\chi^2_{(1).05} = 3.84$ so do not reject H_0. We must conclude that there is no difference in males' and females' preference for either of the two TV shows.

t—F Comparison is analogous to X²—Log Linear Comparison

So, the chi-square test can be used either to test for differences among categories (when there is one sample and two or more categories) or differences among samples (when there are two or more samples and two or more categories). Unlike the ANOVA test which is not limited by the number of independent variables (the only restriction is that there can be only one dependent variable), chi-square is limited to only two variables (which can have any number of levels). As soon as you have more than two variables you can no longer conduct a chi-square test.

Consider the following extension to the problem of old TV shows from above: is there a difference in males' and females' preference for *I Dream of Jeannie* and *Bewitched* as a function of students' class level? We can look at this problem graphically as follows:

278 CHI-SQUARE ANALYSIS

	I Dream of Jeannie			*Bewitched*	
	Male	Female		Male	Female
Freshman			Freshman		
Sophomore			Sophomore		
Junior			Junior		
Senior			Senior		

Figure 11.1 A log-linear problem with nominal-level data and three variables: type of program, gender, and students' class level

There are three variables here: type of program (*I Dream of Jeannie*, *Bewitched*), gender (male, female), and students' class level (freshman, sophomore, junior, senior).

    Recall that with the t-test you could only compare two groups (conditions or levels) of one independent variable. If you added a third (or more) levels to the independent variable, or if you added more independent variables to the problem, you could no longer conduct a t-test and needed to conduct an F-test (in order to control the Type I error, so you would not be conducting multiple t-tests). Exactly the same principle applies to a problem with nominal-level data in which there are more than two variables. To look at all three of these variables via a chi-square, you would need to break this problem down and look at two variables at a time. This is problematic because you would compound the Type I error (chance of being wrong) by conducting multiple statistical tests. Just as the answer to the problem of more than two groups for interval-level data was moving from the t-test to the F-test, the answer to the problem of more than two variables for nominal-level data is to move from a chi-square test to a log-linear analysis. A log-linear analysis is equivalent to the chi-square test for nominal-level data except you are looking at differences among three or more variables simultaneously. As with ANOVA, the concept of main and interaction effects again surfaces with the log-linear analysis.

An Example of Chi-Square Analysis

Let's do an example of chi-square analysis highlighting each step (by now the general process of these steps should be quite familiar to you). Suppose a researcher is interested in whether or not there is a difference in the primary source of political information among students who attend private colleges versus those who attend public universities. He comes up with a list of five possible primary sources of political information: friends, family, political organizations, television, and newspapers. He solicits 100 students from Stanford, a private college, and 100 students from the University of Michigan, a public university, to participate in his study. He asks each of the 200 students which of the five sources is their major source of political information. He obtains the following data:

	Stanford	*U. of Michigan*	
friends	25	35	60
family	20	20	40
political orgs.	30	15	45
television	10	20	30
newspapers	15	10	25
	100	100	200

QUESTION: Is there a difference in private vs. public university students' primary source of political information?

Step 1: State the hypotheses.

Actually, we are asking a research question here: whether two groups of students' sources of political information differ. But, we can state this as alternative and null hypotheses (in words, rather than symbolically, since the symbolic representation of these hypotheses is a bit more difficult to understand clearly):

H_a: Stanford and University of Michigan students differ more than expected by chance in their primary sources of political information.

H_0: There is no difference more than expected by chance in Stanford and University of Michigan students' primary sources of political information.

Step 2: Calculate the theoretical or expected frequencies.

At this point we can determine that this is a chi-square problem. How? Well, first, we are dealing with nominal-level data: frequencies or number of observations per group. Second, there are two variables: type of school and source of political information. This is a multiple-sample chi-square problem; we have two samples, Stanford students and University of Michigan students, and we are looking at differences between these two samples across the five categories of sources of political information.

Note above that the marginal frequencies indicate that there were 100 Stanford students and 100 University of Michigan students, for a grand total of 200 students in the study. A total of 60 students indicated that their major source of political information is their friends, 40 students indicated their major source of political information is their family, and so forth. In computing the theoretical frequencies (what we would expect to observe if there was no difference in the two samples' primary source of political information) we need to maintain these marginal frequencies as they were actually observed:

	Stanford	U. of Michigan	
friends			60
family			40
political orgs.			45
television			30
newspapers			25
	100	100	200

As it happens, this problem involves two equal-size samples, so we can simply take the total number of observations in each category and divide in half. So, for the total of 60 students who indicated their primary source of political information was their friends, 30 of these should be from Stanford and 30 should be from Michigan if there is no difference in the two samples' primary political information source. And so on for the other four categories. Alternatively, we can use the formula for computing theoretical frequencies for the multiple-sample chi-square. For example, for the friends category for Stanford students we would expect:

$$T = \frac{R \times C}{G} = \frac{60 \times 100}{200} = \frac{6,000}{200} = 30$$

So, 30 Stanford students should theoretically be expected to indicate their primary source of political information is their families. If we use the formula (or divide the total observations in each category in half) for each cell, we will obtain the following theoretical table:

	Stanford	U. of Michigan	
friends	30	30	60

family	20	20	40
political orgs.	22.5	22.5	45
television	15	15	30
newspapers	12.5	12.5	25
	100	100	200

Step 3: Calculate the value for χ^2.

If we compare the observed and theoretical values for each of the ten cells using the chi-square formula we obtain the following:

$$\chi^2 = \Sigma \frac{(O-T)^2}{T}$$

$$= \frac{(25-30)^2}{30} + \frac{(35-30)^2}{30} + \frac{(20-20)^2}{20} + \frac{(20-20)^2}{20} + \frac{(30-22.5)^2}{22.5} +$$

$$+ \frac{(15-22.5)^2}{22.5} + \frac{(10-15)^2}{15} + \frac{(20-15)^2}{15} + \frac{(15-12.5)^2}{12.5} + \frac{(10-12.5)^2}{12.5} =$$

$$= 0.833 + 0.833 + 0 + 0 + 2.5 + 2.5 + 1.667 + 1.667 + 0.5 + 0.5 =$$

$$= 11$$

Step 4: Calculate the degrees of freedom.

Recall that we lose one degree of freedom from each of the two variables. The variable type of school has two levels minus one degree of freedom equals one: (C-1) = (2-1) = 1. The variable source of political information has five levels minus one degree of freedom equals four: (R - 1) = (5-1) = 4. We multiply the two sets of degrees of freedom, df = 1 x 4 = 4. The degrees of freedom formula had type of school as the columns (C) and source of political information as the rows (R):

$$df = (C - 1) \times (R - 1) = (2 - 1) \times (5 - 1) = 1 \times 4 = 4$$

Step 5: Look up the critical value at the $p < .05$ level.

Using the χ^2-table we find the following critical value for a χ^2 with 4 degrees of freedom at the .05 significance level:

$$\chi^2_{(4).05} = 9.49$$

Step 6: Compare the calculated χ^2-value to the critical χ^2-value.

Since the calculated value $\chi^2 = 11$ is larger than the critical value $\chi^2_{(4).05} = 9.49$, reject H_0 at the $p < .05$ level and conclude that there is a significant difference between the two samples in their major sources of political information with a Type I error (or chance of being wrong) of .05. Can we do better than a 5% chance of making a mistake?

Step 7: Look up the critical value at the $p < .01$ level.

Using the χ^2-table we find the following critical value for a χ^2 with 4 degrees of freedom at the .01 significance level:

$$\chi^2_{(4).01} = 13.28$$

Step 8: Compare the calculated χ^2-value to the critical χ^2-value.

Since the calculated value of $\chi^2 = 11$ is not larger than the critical value of $\chi^2_{(4).01} = 13.28$ at the .01 significance level we cannot reject H_0 at the more stringent significance level of $p < .01$ to obtain the Type I error of .01. We must live with the 5% error rate.

Step 9: State the final conclusion both statistically and in words.

The final conclusion statistically is to reject H_0 at the $p < .05$ significance level. Conclude that private and public university students differ significantly in their primary sources of political information, with a 5% chance that this is not true. Note that we do not state which sources of political information the two samples differ on. Just as it is with the F-test using an independent variable with more than two levels, so it is with the χ^2 using more than two categories

284 CHI-SQUARE ANALYSIS

(or more than two samples): we cannot state exactly which categories (or samples) differ from which others, only that a difference exists. Since there are only two samples, we know that these two samples differ from each other. However, we do not know whether they differ in the frequency of use of all or only some of the sources of political information. Again, a post-hoc test needs to be conducted to determine where this difference lies. One of the most popular post-hoc tests for the chi-square (and log-linear analysis) is an analog to the Scheffé procedure for the F-test.

Chi-square test: measures the independence between nominal variables;
- asking whether the variable used for row classification is independent of the variable for column classification (Ho) or not (Ha)

← small df
← larger df

- skewed to the right
- concentrates on the positive part (impossible to get a negative x^2 test statistic value)
- the exact shape of the distribution depends on df.

Chapter 11 Problems

1. (*Parametric* or *nonparametric*) tests are based on interval-level measurements and use estimates about populations, whereas (*parametric* or *nonparametric*) tests are based on nominal-level data and do not estimate population characteristics.

2. Indicate the appropriate level(s) of measurement (*nominal, ordinal, interval,* and/or *ratio*) for each of the following (note that more than one may apply):
   a. t-test
   b. mode
   c. standard deviation
   d. chi-square test
   e. correlation (Pearson's r)
   f. median
   g. mean
   h. F-test
   i. linear regression
   j. log-linear analysis
   k. correlation (Spearman's rho)

3. Suppose a legal researcher wishes to determine if people are more likely to become repeat offenders if they commit their first crime before the age of 18, between 18 and 30 years of age, or over 30 years of age. She uses arrest records to obtain a random sample of 50 people who committed their first crime before age 18, 50 people who committed their first crime between 18 and 30, and 50 people who committed their first crime over age 30. She then identifies how many of these people

committed any subsequent crimes (committed additional crimes; did not commit additional crimes). What type of statistical test would the researcher use to analyze her data?

4. What are the two types of chi-square tests and what is the difference between them?

5. In a one-sample or one-way chi-square deign:
   a. What is the formula for chi-square?
   b. How is the theoretical value for a cell calculated?
   c. What is the formula to compute degrees of freedom?
   d. How many variables are present?
   e. Which is analyzed: a difference among categories or a difference among samples?

6. In a multiple-sample or two-way chi-square design:
   a. What is the formula for chi-square?
   b. How is the theoretical value for a cell calculated?
   c. What is the formula to compute degrees of freedom?
   d. How many variables are present?
   e. Which is analyzed: a difference among categories or a difference among samples?

7. For each of the following problems, indicate whether a one-sample chi-square test, a multiple-sample chi-square test, or a log-linear analysis should be conducted. If the test is a chi-square, also provide the total number of degrees of freedom for the statistical test.
   a. Freshmen, sophomores, juniors, and seniors are classified according to their major: economics, mathematics, communication, physics, psychology, sociology, English,

French, German, art, music, law and society, chemistry, or biology.

b. Male and female chief executive officers are categorized according to how much time they spend at work (< 40 hours per week, 40-50 hours per week, > 50 hours per week) and by their yearly income (< $50,000 per year, $50,000-$100,000 per year, > $100,000 per year).

c. Students are divided up according to their letter grade (A, B, C, D, F) in a class.

d. Amusement parks in California (Disneyland, Magic Mountain, Knotts Berry Farm, Sea World) are classified according to their primary ride (roller coasters, water rides, children's rides) and their gross annual profit (< $100 million, $100-$500 million, > $500 million).

e. Eighty subjects are observed initiating a conversation with an elderly person and classified according to their method of beginning the conversation: question about other person, statement about other person, statement about self, statement about environment, question about environment, statement about external event/person, question about external event/person.

f. Fifty men and 50 women are asked to indicate whether they favor the death penalty unconditionally, favor the death penalty only in certain circumstances, or whether they do not favor the death penalty under any conditions.

8. For each of the following chi-square problems, indicate if the chi-square value is significant and at what level.

   a. 4 x 4 contingency table $\chi^2 = 17$
   b. one-sample χ^2 with 6 categories $\chi^2 = 9.86$
   c. 3 x 8 contingency table $\chi^2 = 30.33$

d. 2 x 2 contingency table $\chi^2 = 5.69$
e. one-sample χ^2 with 3 categories $\chi^2 = 10.87$
f. 6 x 6 contingency table $\chi^2 = 36.93$
g. one-sample χ^2 with 10 categories $\chi^2 = 22.32$

9. A ticket agency wants to give away tickets to a Broadway show for promotional purposes. They ask their employees to pick a show currently playing on Broadway that they think will generate the most public interest. The number of employees who picked each Broadway show is as follows:

Phantom	Cats	Oklahoma	My Fair Lady	Miss Saigon
19	15	10	9	9

Is there a significant difference in employee's selection for the Broadway show they think will generate the most public interest? You should conduct a χ^2 analysis and state your answer both statistically and in words.

10. A travel agency surveys college seniors seeking the locations they would most like to travel after obtaining their degrees. The following data are obtained:

Europe	Australia	Africa	China/Far East
40	32	18	23

Is there a significant difference in students' selection of a geographical location? You should conduct a $\chi 2$ analysis and state your answer both statistically and in words.

11. A psychologist and educational researcher is interested in how many unique pieces of information children can remember as a function of the number of distractions in the environment when the information is

being committed to memory. He divides 90 children into three groups (samples) and exposes those in each group to varying levels of environmental distractions: no distractions, television only, television and two people talking. While in their designated environment, children are given a piece of paper with 30 words on it and asked to remember these words. The children are given 10 minutes to learn the words. Later, the researcher asks students to recall as many words from the list as possible. He classifies students into one of three categories: those who remembered 0-10 words, those who remembered 11-20 words, and those who remembered 21-30 words. For the 90 children he obtains the following data:

	0-10 wds	11-20 wds	21-30 wds
No distractions	0	5	25
Television	5	5	20
Television + talking	10	10	10

Is there a significant difference in recall of words among children who learn the words with different levels of environmental distractions (across the three samples: no distractions, television, television and talking)? You should conduct a χ^2 analysis and state your answer both statistically and in words.

12. The university administration is interested in whether students from different class/levels (freshmen, sophomores, juniors, seniors) support stricter eligibility requirements to remain enrolled at the university. They ask 100 randomly selected students from each class whether they support stricter standards. The following data is obtained:

	Support	Do Not Support
Freshmen	25	75
Sophomores	25	75

APPENDICES

Critical Values of *T*

For any given df, the table shows the values of *t* corresponding to various levels of probability. The obtained *t* is significant at a given level if it is equal to or *greater than* the value shown in the table.

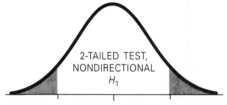

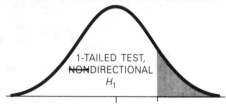

	Level of significance		
df	.05	.01	.001
1	12.706	63.657	636.619
2	4.303	9.925	31.598
3	3.182	5.841	12.941
4	2.776	4.604	8.610
5	2.571	4.032	6.859
6	2.447	3.707	5.959
7	2.365	3.499	5.405
8	2.306	3.355	5.041
9	2.262	3.250	4.781
10	2.228	3.169	4.587
11	2.201	3.106	4.437
12	2.179	3.055	4.318
13	2.160	3.012	4.221
14	2.145	2.977	4.140
15	2.131	2.947	4.073
16	2.120	2.921	4.015
17	2.110	2.898	3.965
18	2.101	2.878	3.922
19	2.093	2.861	3.883
20	2.086	2.845	3.850
21	2.080	2.831	3.819
22	2.074	2.819	3.792
23	2.069	2.807	3.767
24	2.064	2.797	3.745
25	2.060	2.787	3.725
26	2.056	2.779	3.707
27	2.052	2.771	3.690
28	2.048	2.763	3.674
29	2.045	2.756	3.659
30	2.042	2.750	3.646
40	2.021	2.704	3.551
60	2.000	2.660	3.460
120	1.980	2.617	3.373
∞	1.960	2.576	3.291

	Level of significance		
df	.05	.01	.005
1	6.314	31.821	63.657
2	2.920	6.965	9.925
3	2.353	4.541	5.841
4	2.132	3.747	4.604
5	2.015	3.365	4.032
6	1.943	3.143	3.707
7	1.895	2.998	3.499
8	1.860	2.896	3.355
9	1.833	2.821	3.250
10	1.812	2.764	3.169
11	1.796	2.718	3.106
12	1.782	2.681	3.055
13	1.771	2.650	3.012
14	1.761	2.624	2.977
15	1.753	2.602	2.947
16	1.746	2.583	2.921
17	1.740	2.567	2.898
18	1.734	2.552	2.878
19	1.729	2.539	2.861
20	1.725	2.528	2.845
21	1.721	2.518	2.831
22	1.717	2.508	2.819
23	1.714	2.500	2.807
24	1.711	2.492	2.797
25	1.708	2.485	2.787
26	1.706	2.479	2.779
27	1.703	2.473	2.771
28	1.701	2.467	2.763
29	1.699	2.462	2.756
30	1.697	2.457	2.750
40	1.684	2.423	2.704
60	1.671	2.390	2.660
120	1.658	2.358	2.617
∞	1.645	2.326	2.576

Source: Table C is taken from Table III (page 46) of Fisher and Yates, *Statistical Tables for Biological, Agricultural, and Medical Research,* 6th ed., published by Longman Group Ltd., 1974. London (previously published by Oliver and Boyd, Edinburgh), and by permission of the authors and publishers.

Critical Values of F

The obtained F is significant at a given level if it is equal to or *greater than* the value shown in the table. 0.05 (top row in each pair) and 0.01 (bottom row in each pair) points for the distribution of F.

The values shown are the right tail of the distribution obtained by dividing the larger variance estimate by the smaller variance estimate. To find the complementary left or lower tail for a given df and α-level, reverse the degrees of freedom and find the reciprocal of that value in the F-table. For example, the value cutting off the top 5% of the area for 7 and 12 df is 2.85. To find the cutoff point of the bottom 5% of the area, find the tabled value of the $\alpha = 0.05$ level for 12 and 7 df. This is found to be 3.57. The reciprocal is $1/3.57 = 0.28$. Thus 5% of the area falls *at or below an* $F = 0.28$.

Source: G.W. Snedecor and W. G. Cochran, *Statistical Methods*, 8th ed. © 1989 by Iowa State University Press, Ames, Iowa. Reprinted by permission.

	1	2	3	4	5	6	7	8	9	10	11	12	14	16	20	24	30	40	50	75	100	200	500	∞
1	161 / 4052	200 / 4999	216 / 5403	225 / 5625	230 / 5764	234 / 5859	237 / 5928	239 / 5981	241 / 6022	242 / 6056	243 / 6082	244 / 6106	245 / 6142	246 / 6169	248 / 6208	249 / 6234	250 / 6258	251 / 6286	252 / 6302	253 / 6323	253 / 6334	254 / 6352	254 / 6361	254 / 6366
2	18.51 / 98.49	19.00 / 99.01	19.16 / 99.17	19.25 / 99.25	19.30 / 99.30	19.33 / 99.33	19.36 / 99.34	19.37 / 99.36	19.38 / 99.38	19.39 / 99.40	19.40 / 99.41	19.41 / 99.42	19.42 / 99.43	19.43 / 99.44	19.44 / 99.45	19.45 / 99.46	19.46 / 99.47	19.47 / 99.48	19.47 / 99.48	19.48 / 99.49	19.49 / 99.49	19.49 / 99.49	19.50 / 99.50	19.50 / 99.50
3	10.13 / 34.12	9.55 / 30.81	9.28 / 29.46	9.12 / 28.71	9.01 / 28.24	8.94 / 27.91	8.88 / 27.67	8.84 / 27.49	8.81 / 27.34	8.78 / 27.23	8.76 / 27.13	8.74 / 27.05	8.71 / 26.92	8.69 / 26.83	8.66 / 26.69	8.64 / 26.60	8.62 / 26.50	8.60 / 26.41	8.58 / 26.30	8.57 / 26.27	8.56 / 26.23	8.54 / 26.18	8.54 / 26.14	8.53 / 26.12
4	7.71 / 21.20	6.94 / 18.00	6.59 / 16.69	6.39 / 15.98	6.26 / 15.52	6.16 / 15.21	6.09 / 14.98	6.04 / 14.80	6.00 / 14.66	5.96 / 14.54	5.93 / 14.45	5.91 / 14.37	5.87 / 14.24	5.84 / 14.15	5.80 / 14.02	5.77 / 13.93	5.74 / 13.83	5.71 / 13.74	5.70 / 13.69	5.68 / 13.61	5.66 / 13.57	5.65 / 13.52	5.64 / 13.48	5.63 / 13.46
5	6.61 / 16.26	5.79 / 13.27	5.41 / 12.06	5.19 / 11.39	5.05 / 10.97	4.95 / 10.67	4.88 / 10.45	4.82 / 10.27	4.78 / 10.15	4.74 / 10.05	4.70 / 9.96	4.68 / 9.89	4.64 / 9.77	4.60 / 9.68	4.56 / 9.55	4.53 / 9.47	4.50 / 9.38	4.46 / 9.29	4.44 / 9.24	4.42 / 9.17	4.40 / 9.13	4.38 / 9.07	4.37 / 9.04	4.36 / 9.02
6	5.99 / 13.74	5.14 / 10.92	4.76 / 9.78	4.53 / 9.15	4.39 / 8.75	4.28 / 8.47	4.21 / 8.26	4.15 / 8.10	4.10 / 7.98	4.06 / 7.87	4.03 / 7.79	4.00 / 7.72	3.96 / 7.60	3.92 / 7.52	3.87 / 7.39	3.84 / 7.31	3.81 / 7.23	3.77 / 7.14	3.75 / 7.09	3.72 / 7.02	3.71 / 6.99	3.69 / 6.94	3.68 / 6.90	3.67 / 6.88
7	5.59 / 12.25	4.74 / 9.55	4.35 / 8.45	4.12 / 7.85	3.97 / 7.46	3.87 / 7.19	3.79 / 7.00	3.73 / 6.84	3.68 / 6.71	3.63 / 6.62	3.60 / 6.54	3.57 / 6.47	3.52 / 6.35	3.49 / 6.27	3.44 / 6.15	3.41 / 6.07	3.38 / 5.98	3.34 / 5.90	3.32 / 5.85	3.29 / 5.78	3.28 / 5.75	3.25 / 5.70	3.24 / 5.67	3.23 / 5.65
8	5.32 / 11.26	4.46 / 8.65	4.07 / 7.59	3.84 / 7.01	3.69 / 6.63	3.58 / 6.37	3.50 / 6.19	3.44 / 6.03	3.39 / 5.91	3.34 / 5.82	3.31 / 5.74	3.28 / 5.67	3.23 / 5.56	3.20 / 5.48	3.15 / 5.36	3.12 / 5.28	3.08 / 5.20	3.05 / 5.11	3.03 / 5.06	3.00 / 5.00	2.98 / 4.96	2.96 / 4.91	2.94 / 4.88	2.93 / 4.86
9	5.12 / 10.56	4.26 / 8.02	3.86 / 6.99	3.63 / 6.42	3.48 / 6.06	3.37 / 5.80	3.29 / 5.62	3.23 / 5.47	3.18 / 5.35	3.13 / 5.26	3.10 / 5.18	3.07 / 5.11	3.02 / 5.00	2.98 / 4.92	2.93 / 4.80	2.90 / 4.73	2.86 / 4.64	2.82 / 4.56	2.80 / 4.51	2.77 / 4.45	2.76 / 4.41	2.73 / 4.36	2.72 / 4.33	2.71 / 4.31
10	4.96 / 10.04	4.10 / 7.56	3.71 / 6.55	3.48 / 5.99	3.33 / 5.64	3.22 / 5.39	3.14 / 5.21	3.07 / 5.06	3.02 / 4.95	2.97 / 4.85	2.94 / 4.78	2.91 / 4.71	2.86 / 4.60	2.82 / 4.52	2.77 / 4.41	2.74 / 4.33	2.70 / 4.25	2.67 / 4.17	2.64 / 4.12	2.61 / 4.05	2.59 / 4.01	2.56 / 3.96	2.55 / 3.93	2.54 / 3.91
11	4.84 / 9.65	3.98 / 7.20	3.59 / 6.22	3.36 / 5.67	3.20 / 5.32	3.09 / 5.07	3.01 / 4.88	2.95 / 4.74	2.90 / 4.63	2.86 / 4.54	2.82 / 4.46	2.79 / 4.40	2.74 / 4.29	2.70 / 4.21	2.65 / 4.10	2.61 / 4.02	2.57 / 3.94	2.53 / 3.86	2.50 / 3.80	2.47 / 3.74	2.45 / 3.70	2.42 / 3.66	2.41 / 3.62	2.40 / 3.60
12	4.75 / 9.33	3.88 / 6.93	3.49 / 5.95	3.26 / 5.41	3.11 / 5.06	3.00 / 4.82	2.92 / 4.65	2.85 / 4.50	2.80 / 4.39	2.76 / 4.30	2.72 / 4.22	2.69 / 4.16	2.64 / 4.05	2.60 / 3.98	2.54 / 3.86	2.50 / 3.78	2.46 / 3.70	2.42 / 3.61	2.40 / 3.56	2.36 / 3.49	2.35 / 3.46	2.32 / 3.41	2.31 / 3.38	2.30 / 3.36
13	4.67 / 9.07	3.80 / 6.70	3.41 / 5.74	3.18 / 5.20	3.02 / 4.86	2.92 / 4.62	2.84 / 4.44	2.77 / 4.30	2.72 / 4.19	2.67 / 4.10	2.63 / 4.02	2.60 / 3.96	2.55 / 3.85	2.51 / 3.78	2.46 / 3.67	2.42 / 3.59	2.38 / 3.51	2.34 / 3.42	2.32 / 3.37	2.28 / 3.30	2.26 / 3.27	2.24 / 3.21	2.22 / 3.18	2.21 / 3.16
14	4.60 / 8.86	3.74 / 6.51	3.34 / 5.56	3.11 / 5.03	2.96 / 4.69	2.85 / 4.46	2.77 / 4.28	2.70 / 4.14	2.65 / 4.03	2.60 / 3.94	2.56 / 3.86	2.53 / 3.80	2.48 / 3.70	2.44 / 3.62	2.39 / 3.51	2.35 / 3.43	2.31 / 3.34	2.27 / 3.26	2.24 / 3.21	2.21 / 3.14	2.19 / 3.11	2.16 / 3.06	2.14 / 3.02	2.13 / 3.00
15	4.54 / 8.68	3.68 / 6.36	3.29 / 5.42	3.06 / 4.89	2.90 / 4.56	2.79 / 4.32	2.70 / 4.14	2.64 / 4.00	2.59 / 3.89	2.55 / 3.80	2.51 / 3.73	2.48 / 3.67	2.43 / 3.56	2.39 / 3.48	2.33 / 3.36	2.29 / 3.29	2.25 / 3.20	2.21 / 3.12	2.18 / 3.07	2.15 / 3.00	2.12 / 2.97	2.10 / 2.92	2.08 / 2.89	2.07 / 2.87

Degrees of freedom for numerator (columns); Degrees of freedom for denominator (rows)

(Continued)

df																					
16	4.49 8.53	3.63 6.23	3.24 5.29	3.01 4.77	2.85 4.44	2.74 4.20	2.66 4.03	2.59 3.89	2.54 3.78	2.49 3.69	2.45 3.61	2.42 3.55	2.37 3.45	2.33 3.37	2.28 3.25	2.24 3.18	2.20 3.10	2.16 3.01	2.13 2.96	2.09 2.89	2.07 2.86	2.04 2.80	2.02 2.77	2.01 2.75
17	4.45 8.40	3.59 6.11	3.20 5.18	2.96 4.67	2.81 4.34	2.70 4.10	2.62 3.93	2.55 3.79	2.50 3.68	2.45 3.59	2.41 3.52	2.38 3.45	2.33 3.35	2.29 3.27	2.23 3.16	2.19 3.08	2.15 3.00	2.11 2.92	2.08 2.86	2.04 2.79	2.02 2.76	1.99 2.70	1.97 2.67	1.96 2.65
18	4.41 8.28	3.55 6.01	3.16 5.09	2.93 4.58	2.77 4.25	2.66 4.01	2.58 3.85	2.51 3.71	2.46 3.60	2.41 3.51	2.37 3.44	2.34 3.37	2.29 3.27	2.25 3.19	2.19 3.07	2.15 3.00	2.11 2.91	2.07 2.83	2.04 2.78	2.00 2.71	1.98 2.68	1.95 2.62	1.93 2.59	1.92 2.57
19	4.38 8.18	3.52 5.93	3.13 5.01	2.90 4.50	2.74 4.17	2.63 3.94	2.55 3.77	2.48 3.63	2.43 3.52	2.38 3.43	2.34 3.36	2.31 3.30	2.26 3.19	2.21 3.12	2.15 3.00	2.11 2.92	2.07 2.84	2.02 2.76	2.00 2.70	1.96 2.63	1.94 2.60	1.91 2.54	1.90 2.51	1.88 2.49
20	4.35 8.10	3.49 5.85	3.10 4.94	2.87 4.43	2.71 4.10	2.60 3.87	2.52 3.71	2.45 3.56	2.40 3.45	2.35 3.37	2.31 3.30	2.28 3.23	2.23 3.13	2.18 3.05	2.12 2.94	2.08 2.86	2.04 2.77	1.99 2.69	1.96 2.63	1.92 2.56	1.90 2.53	1.87 2.47	1.85 2.44	1.84 2.42
21	4.32 8.02	3.47 5.78	3.07 4.87	2.84 4.37	2.68 4.04	2.57 3.81	2.49 3.65	2.42 3.51	2.37 3.40	2.32 3.31	2.28 3.24	2.25 3.17	2.20 3.07	2.15 2.99	2.09 2.88	2.05 2.80	2.00 2.72	1.96 2.63	1.93 2.58	1.89 2.51	1.87 2.47	1.84 2.42	1.82 2.38	1.81 2.36
22	4.30 7.94	3.44 5.72	3.05 4.82	2.82 4.31	2.66 3.99	2.55 3.76	2.47 3.59	2.40 3.45	2.35 3.35	2.30 3.26	2.26 3.18	2.23 3.12	2.18 3.02	2.13 2.94	2.07 2.83	2.03 2.75	1.98 2.67	1.93 2.58	1.91 2.53	1.87 2.46	1.84 2.42	1.81 2.37	1.80 2.33	1.78 2.31
23	4.28 7.88	3.42 5.66	3.03 4.76	2.80 4.26	2.64 3.94	2.53 3.71	2.45 3.54	2.38 3.41	2.32 3.30	2.28 3.21	2.24 3.14	2.20 3.07	2.14 2.97	2.10 2.89	2.04 2.78	2.00 2.70	1.96 2.62	1.91 2.53	1.88 2.48	1.84 2.41	1.82 2.37	1.79 2.32	1.77 2.28	1.76 2.26
24	4.26 7.82	3.40 5.61	3.01 4.72	2.78 4.22	2.62 3.90	2.51 3.67	2.43 3.50	2.36 3.36	2.30 3.25	2.26 3.17	2.22 3.09	2.18 3.03	2.13 2.93	2.09 2.85	2.02 2.74	1.98 2.66	1.94 2.58	1.89 2.49	1.86 2.44	1.82 2.36	1.80 2.33	1.76 2.27	1.74 2.23	1.73 2.21
25	4.24 7.77	3.38 5.57	2.99 4.68	2.76 4.18	2.60 3.86	2.49 3.63	2.41 3.46	2.34 3.32	2.28 3.21	2.24 3.13	2.20 3.05	2.16 2.99	2.11 2.89	2.06 2.81	2.00 2.70	1.96 2.62	1.92 2.54	1.87 2.45	1.84 2.40	1.80 2.32	1.77 2.29	1.74 2.23	1.72 2.19	1.71 2.17
26	4.22 7.72	3.37 5.53	2.96 4.64	2.74 4.14	2.59 3.82	2.47 3.59	2.39 3.42	2.32 3.29	2.27 3.17	2.22 3.09	2.18 3.02	2.15 2.96	2.10 2.86	2.05 2.77	1.99 2.66	1.95 2.58	1.90 2.50	1.85 2.41	1.82 2.36	1.78 2.28	1.76 2.25	1.72 2.19	1.70 2.15	1.69 2.13
27	4.21 7.68	3.35 5.49	2.95 4.60	2.73 4.11	2.57 3.79	2.46 3.56	2.37 3.39	2.30 3.26	2.25 3.14	2.20 3.06	2.16 2.98	2.13 2.93	2.08 2.83	2.03 2.74	1.97 2.63	1.93 2.55	1.88 2.47	1.84 2.38	1.80 2.33	1.76 2.25	1.74 2.21	1.71 2.16	1.68 2.12	1.67 2.10
28	4.20 7.64	3.34 5.45	2.93 4.57	2.71 4.07	2.56 3.76	2.44 3.53	2.36 3.36	2.29 3.23	2.24 3.11	2.19 3.03	2.15 2.95	2.12 2.90	2.06 2.80	2.02 2.71	1.96 2.60	1.91 2.52	1.87 2.44	1.81 2.35	1.78 2.30	1.75 2.22	1.72 2.18	1.69 2.13	1.67 2.09	1.65 2.06
29	4.18 7.60	3.33 5.52	2.93 4.54	2.70 4.04	2.54 3.73	2.43 3.50	2.35 3.32	2.28 3.20	2.22 3.08	2.18 3.00	2.14 2.92	2.10 2.87	2.05 2.77	2.00 2.68	1.94 2.57	1.90 2.49	1.85 2.41	1.80 2.32	1.77 2.27	1.73 2.19	1.71 2.15	1.68 2.10	1.65 2.06	1.64 2.03
30	4.17 7.56	3.32 5.39	2.92 4.51	2.69 4.02	2.53 3.70	2.42 3.47	2.34 3.30	2.27 3.17	2.21 3.06	2.16 2.98	2.12 2.90	2.09 2.84	2.04 2.74	1.99 2.66	1.93 2.55	1.89 2.47	1.84 2.38	1.79 2.29	1.76 2.24	1.72 2.16	1.69 2.13	1.66 2.07	1.64 2.03	1.62 2.01
32	4.15 7.50	3.30 5.34	2.90 4.46	2.67 3.97	2.51 3.66	2.40 3.42	2.32 3.25	2.25 3.12	2.19 3.01	2.14 2.94	2.10 2.86	2.07 2.80	2.02 2.70	1.97 2.62	1.91 2.51	1.86 2.42	1.82 2.34	1.76 2.25	1.74 2.20	1.69 2.12	1.67 2.08	1.64 2.02	1.61 1.98	1.59 1.96
34	4.13 7.44	3.28 5.29	2.88 4.42	2.65 3.93	2.49 3.61	2.38 3.38	2.30 3.21	2.23 3.08	2.17 2.97	2.12 2.89	2.08 2.82	2.05 2.76	2.00 2.66	1.95 2.58	1.89 2.47	1.84 2.38	1.80 2.30	1.74 2.21	1.71 2.15	1.67 2.08	1.64 2.04	1.61 1.98	1.59 1.94	1.57 1.91

Degrees of freedom for denominator

df denom \ df num	1	2	3	4	5	6	7	8	9	10	11	12	14	16	20	24	30	40	50	75	100	200	500	∞
36	4.11 / 7.39	3.26 / 5.25	2.86 / 4.38	2.63 / 3.89	2.48 / 3.58	2.36 / 3.35	2.28 / 3.18	2.21 / 3.04	2.15 / 2.94	2.10 / 2.86	2.06 / 2.78	2.03 / 2.72	1.89 / 2.62	1.93 / 2.54	1.87 / 2.43	1.82 / 2.35	1.78 / 2.26	1.72 / 2.17	1.69 / 2.12	1.65 / 2.04	1.62 / 2.00	1.59 / 1.94	1.56 / 1.90	1.55 / 1.87
38	4.10 / 7.35	3.25 / 5.21	2.85 / 4.34	2.62 / 3.86	2.46 / 3.54	2.35 / 3.32	2.26 / 3.15	2.19 / 3.02	2.14 / 2.91	2.09 / 2.82	2.05 / 2.75	2.02 / 2.69	1.96 / 2.59	1.92 / 2.51	1.85 / 2.40	1.80 / 2.32	1.76 / 2.22	1.71 / 2.14	1.67 / 2.08	1.63 / 2.00	1.60 / 1.97	1.57 / 1.90	1.54 / 1.86	1.53 / 1.84
40	4.08 / 7.31	3.23 / 5.18	2.84 / 4.31	2.61 / 3.83	2.45 / 3.51	2.34 / 3.29	2.25 / 3.12	2.18 / 2.99	2.12 / 2.88	2.07 / 2.80	2.04 / 2.73	2.00 / 2.66	1.95 / 2.56	1.90 / 2.49	1.84 / 2.37	1.79 / 2.29	1.74 / 2.20	1.69 / 2.11	1.66 / 2.05	1.61 / 1.97	1.59 / 1.94	1.55 / 1.88	1.53 / 1.84	1.51 / 1.81
42	4.07 / 7.27	3.22 / 5.15	2.83 / 4.29	2.59 / 3.80	2.44 / 3.49	2.32 / 3.26	2.24 / 3.10	2.17 / 2.96	2.11 / 2.86	2.06 / 2.77	2.02 / 2.70	1.99 / 2.64	1.94 / 2.54	1.89 / 2.46	1.82 / 2.35	1.78 / 2.26	1.73 / 2.17	1.68 / 2.08	1.64 / 2.02	1.60 / 1.94	1.57 / 1.91	1.54 / 1.85	1.51 / 1.80	1.49 / 1.78
44	4.06 / 7.24	3.21 / 5.12	2.82 / 4.26	2.58 / 3.78	2.43 / 3.46	2.31 / 3.24	2.23 / 3.07	2.16 / 2.94	2.10 / 2.84	2.05 / 2.75	2.01 / 2.68	1.98 / 2.62	1.92 / 2.52	1.88 / 2.44	1.81 / 2.32	1.76 / 2.24	1.72 / 2.15	1.66 / 2.06	1.63 / 2.00	1.58 / 1.92	1.56 / 1.88	1.52 / 1.82	1.50 / 1.78	1.48 / 1.75
46	4.05 / 7.21	3.20 / 5.10	2.81 / 4.24	2.57 / 3.76	2.42 / 3.44	2.30 / 3.22	2.22 / 3.05	2.14 / 2.92	2.09 / 2.82	2.04 / 2.73	2.00 / 2.66	1.97 / 2.60	1.91 / 2.50	1.87 / 2.42	1.80 / 2.30	1.75 / 2.22	1.71 / 2.13	1.65 / 2.04	1.62 / 1.98	1.57 / 1.90	1.54 / 1.86	1.51 / 1.80	1.48 / 1.76	1.46 / 1.72
48	4.04 / 7.19	3.19 / 5.08	2.80 / 4.22	2.56 / 3.74	2.41 / 3.42	2.30 / 3.20	2.21 / 3.04	2.14 / 2.90	2.08 / 2.80	2.03 / 2.71	1.99 / 2.64	1.96 / 2.58	1.90 / 2.48	1.86 / 2.40	1.79 / 2.28	1.74 / 2.20	1.70 / 2.11	1.64 / 2.02	1.61 / 1.96	1.56 / 1.88	1.53 / 1.84	1.50 / 1.78	1.47 / 1.73	1.45 / 1.70
50	4.03 / 7.17	3.18 / 5.06	2.79 / 4.20	2.56 / 3.72	2.40 / 3.41	2.29 / 3.18	2.20 / 3.02	2.13 / 2.88	2.07 / 2.78	2.02 / 2.70	1.98 / 2.62	1.95 / 2.56	1.90 / 2.46	1.85 / 2.39	1.78 / 2.26	1.74 / 2.18	1.69 / 2.10	1.63 / 2.00	1.60 / 1.94	1.55 / 1.86	1.52 / 1.82	1.48 / 1.76	1.46 / 1.71	1.44 / 1.68
55	4.02 / 7.12	3.17 / 5.01	2.78 / 4.16	2.54 / 3.68	2.38 / 3.37	2.27 / 3.15	2.18 / 2.98	2.11 / 2.85	2.05 / 2.75	2.00 / 2.66	1.97 / 2.59	1.93 / 2.53	1.88 / 2.43	1.83 / 2.35	1.76 / 2.23	1.72 / 2.15	1.67 / 2.06	1.61 / 1.96	1.58 / 1.90	1.52 / 1.82	1.50 / 1.78	1.46 / 1.71	1.43 / 1.66	1.41 / 1.64
60	4.00 / 7.08	3.15 / 4.98	2.76 / 4.13	2.52 / 3.65	2.37 / 3.34	2.25 / 3.12	2.17 / 2.95	2.10 / 2.82	2.04 / 2.72	1.99 / 2.63	1.95 / 2.56	1.92 / 2.50	1.86 / 2.40	1.81 / 2.32	1.75 / 2.20	1.70 / 2.12	1.65 / 2.03	1.59 / 1.93	1.56 / 1.87	1.50 / 1.79	1.48 / 1.74	1.44 / 1.68	1.41 / 1.63	1.39 / 1.60
65	3.99 / 7.04	3.14 / 4.95	2.75 / 4.10	2.51 / 3.62	2.36 / 3.31	2.24 / 3.09	2.15 / 2.93	2.08 / 2.79	2.02 / 2.70	1.98 / 2.61	1.94 / 2.54	1.90 / 2.47	1.85 / 2.37	1.80 / 2.30	1.73 / 2.18	1.68 / 2.09	1.63 / 2.00	1.57 / 1.90	1.54 / 1.84	1.49 / 1.76	1.46 / 1.71	1.42 / 1.64	1.39 / 1.60	1.37 / 1.56
70	3.98 / 7.01	3.13 / 4.92	2.74 / 4.08	2.50 / 3.60	2.35 / 3.29	2.32 / 3.07	2.14 / 2.91	2.07 / 2.77	2.01 / 2.67	1.97 / 2.59	1.93 / 2.51	1.89 / 2.45	1.84 / 2.35	1.79 / 2.28	1.72 / 2.15	1.67 / 2.07	1.62 / 1.98	1.56 / 1.88	1.53 / 1.82	1.47 / 1.74	1.45 / 1.69	1.40 / 1.62	1.37 / 1.56	1.35 / 1.53
80	3.96 / 6.96	3.11 / 4.88	2.72 / 4.04	2.48 / 3.56	2.33 / 3.25	2.21 / 3.04	2.12 / 2.87	2.05 / 2.74	1.99 / 2.64	1.95 / 2.55	1.91 / 2.48	1.88 / 2.41	1.82 / 2.32	1.77 / 2.24	1.70 / 2.11	1.65 / 2.03	1.60 / 1.94	1.54 / 1.84	1.51 / 1.78	1.45 / 1.70	1.42 / 1.65	1.38 / 1.57	1.35 / 1.52	1.32 / 1.49
100	3.94 / 6.90	3.09 / 4.82	2.70 / 3.98	2.46 / 3.51	2.30 / 3.20	2.19 / 2.99	2.10 / 2.82	2.03 / 2.69	1.97 / 2.59	1.92 / 2.51	1.88 / 2.43	1.85 / 2.36	1.79 / 2.26	1.75 / 2.19	1.68 / 2.06	1.63 / 1.98	1.57 / 1.89	1.51 / 1.79	1.48 / 1.73	1.42 / 1.64	1.39 / 1.59	1.34 / 1.51	1.30 / 1.46	1.28 / 1.43
125	3.92 / 6.84	3.07 / 4.78	2.68 / 3.94	2.44 / 3.47	2.29 / 3.17	2.17 / 2.95	2.08 / 2.79	2.01 / 2.65	1.95 / 2.56	1.90 / 2.47	1.86 / 2.40	1.83 / 2.33	1.77 / 2.23	1.72 / 2.15	1.65 / 2.03	1.60 / 1.94	1.55 / 1.85	1.49 / 1.75	1.45 / 1.68	1.39 / 1.59	1.36 / 1.54	1.31 / 1.46	1.27 / 1.40	1.25 / 1.37
150	3.91 / 6.81	3.06 / 4.75	2.67 / 3.91	2.43 / 3.44	2.27 / 3.13	2.16 / 2.92	2.07 / 2.76	2.00 / 2.62	1.94 / 2.53	1.89 / 2.44	1.85 / 2.37	1.82 / 2.30	1.76 / 2.20	1.71 / 2.12	1.64 / 2.00	1.59 / 1.91	1.54 / 1.83	1.47 / 1.72	1.44 / 1.66	1.37 / 1.56	1.34 / 1.51	1.29 / 1.43	1.25 / 1.37	1.22 / 1.33
200	3.89 / 6.76	3.04 / 4.71	2.65 / 3.88	2.41 / 3.41	2.26 / 3.11	2.14 / 2.90	2.05 / 2.73	1.98 / 2.60	1.92 / 2.50	1.87 / 2.41	1.83 / 2.34	1.80 / 2.28	1.74 / 1.17	1.69 / 2.09	1.62 / 1.97	1.57 / 1.88	1.52 / 1.79	1.45 / 1.69	1.42 / 1.62	1.35 / 1.53	1.32 / 1.48	1.26 / 1.39	1.22 / 1.33	1.19 / 1.28
400	3.86 / 6.70	3.02 / 4.66	2.62 / 3.83	2.39 / 3.36	2.23 / 3.06	2.12 / 2.85	2.03 / 2.69	1.96 / 2.55	1.90 / 2.46	1.85 / 2.37	1.81 / 2.29	1.78 / 2.23	1.72 / 2.12	1.67 / 2.04	1.60 / 1.92	1.54 / 1.84	1.49 / 1.74	1.42 / 1.64	1.38 / 1.57	1.32 / 1.47	1.28 / 1.42	1.22 / 1.32	1.16 / 1.24	1.13 / 1.19
1000	3.85 / 6.66	3.00 / 4.62	2.61 / 3.80	2.38 / 3.34	2.22 / 3.04	2.10 / 2.82	2.02 / 2.66	1.95 / 2.53	1.89 / 2.43	1.84 / 2.34	1.80 / 2.26	1.76 / 2.20	1.70 / 2.09	1.65 / 2.01	1.58 / 1.89	1.53 / 1.81	1.47 / 1.71	1.41 / 1.61	1.36 / 1.54	1.30 / 1.44	1.26 / 1.38	1.19 / 1.28	1.13 / 1.19	1.08 / 1.11
∞	3.84 / 6.64	2.99 / 4.60	2.60 / 3.78	2.37 / 3.32	2.21 / 3.02	2.09 / 2.80	2.01 / 2.64	1.94 / 2.51	1.88 / 2.41	1.83 / 2.32	1.79 / 2.24	1.75 / 2.18	1.69 / 2.07	1.64 / 1.99	1.57 / 1.87	1.52 / 1.79	1.46 / 1.69	1.40 / 1.59	1.35 / 1.52	1.28 / 1.41	1.24 / 1.36	1.17 / 1.25	1.11 / 1.15	1.00 / 1.00

Critical Values of r

n	$\cdot 1$	$\cdot 05$	$\cdot 02$	$\cdot 01$	$\cdot 001$
1	·98769	·99692	·999507	·999877	·9999988
2	·90000	·95000	·98000	·99000	·99900
3	·8054	·8783	·93433	·95873	·99116
4	·7293	·8114	·8822	·91720	·97406
5	·6694	·7545	·8329	·8745	·95074
6	·6215	·7067	·7887	·8343	·92493
7	·5822	·6664	·7498	·7977	·8982
8	·5494	·6319	·7155	·7646	·8721
9	·5214	·6021	·6851	·7348	·8471
10	·4973	·5760	·6581	·7079	·8233
11	·4762	·5529	·6339	·6835	·8010
12	·4575	·5324	·6120	·6614	·7800
13	·4409	·5139	·5923	·6411	·7603
14	·4259	·4973	·5742	·6226	·7420
15	·4124	·4821	·5577	·6055	·7246

n	$\cdot 1$	$\cdot 05$	$\cdot 02$	$\cdot 01$	$\cdot 001$
16	·4000	·4683	·5425	·5897	·7084
17	·3887	·4555	·5285	·5751	·6932
18	·3783	·4438	·5155	·5614	·6787
19	·3687	·4329	·5034	·5487	·6652
20	·3598	·4227	·4921	·5368	·6524
25	·3233	·3809	·4451	·4869	·5974
30	·2960	·3494	·4093	·4487	·5541
35	·2746	·3246	·3810	·4182	·5189
40	·2573	·3044	·3578	·3932	·4896
45	·2428	·2875	·3384	·3721	·4648
50	·2306	·2732	·3218	·3541	·4433
60	·2108	·2500	·2948	·3248	·4078
70	·1954	·2319	·2737	·3017	·3799
80	·1829	·2172	·2565	·2830	·3568
90	·1726	·2050	·2422	·2673	·3375
100	·1638	·1946	·2301	·2540	·3211

Source: Table C is taken from Table III (**page 46**) of Fisher and Yates, *Statistical Tables for Biological, Agricultural, and Medical Research*, 6th ed., published by Longman Group Ltd., 1974, London (previously published by Oliver and Boyd, Edinburgh), and by permission of the authors and publishers.

Critical Values of χ^2

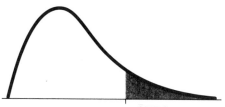

χ^2 CRITICAL VALUE

Degrees of freedom df	.10	.05	.02	.01
1	2.706	3.841	5.412	6.635
2	4.605	5.991	7.824	9.210
3	6.251	7.815	9.837	11.341
4	7.779	9.488	11.668	13.277
5	9.236	11.070	13.388	15.086
6	10.645	12.592	15.033	16.812
7	12.017	14.067	16.622	18.475
8	13.362	15.507	18.168	20.090
9	14.684	16.919	19.679	21.666
10	15.987	18.307	21.161	23.209
11	17.275	19.675	22.618	24.725
12	18.549	21.026	24.054	26.217
13	19.812	22.362	25.472	27.688
14	21.064	23.685	26.873	29.141
15	22.307	24.996	28.259	30.578
16	23.542	26.296	29.633	32.000
17	24.769	27.587	30.995	33.409
18	25.989	28.869	32.346	34.805
19	27.204	30.144	33.687	36.191
20	28.412	31.410	35.020	37.566
21	29.615	32.671	36.343	38.932
22	30.813	33.924	37.659	40.289
23	32.007	35.172	38.968	41.638
24	33.196	36.415	40.270	42.980
25	34.382	37.652	41.566	44.314
26	35.563	38.885	42.856	45.642
27	36.741	40.113	44.140	46.963
28	37.916	41.337	45.419	48.278
29	39.087	42.557	46.693	49.588
30	40.256	43.773	47.962	50.892

Source: Ronald A. Fisher, *Statistical Methods for Research Workers*, 14th ed., Table III, p. 113. Copyright © 1970 by University of Adelaide. Reprinted by permission of Luminis Pty., Ltd., Adelaide, South Australia.

Table A Normal curve areas

Standard normal probability in right-hand tail (for negative values of z, areas are found by symmetry)

remember this is only for one side so 1/2 the actual

SECOND DECIMAL PLACE OF z

z	.00	.01	.02	.03	.04	.05	.06	.07	.08	.09
0.0	.5000	.4960	.4920	.4880	.4840	.4801	.4761	.4721	.4681	.4641
0.1	.4602	.4562	.4522	.4483	.4443	.4404	.4364	.4325	.4286	.4247
0.2	.4207	.4168	.4129	.4090	.4052	.4013	.3974	.3936	.3897	.3859
0.3	.3821	.3783	.3745	.3707	.3669	.3632	.3594	.3557	.3520	.3483
0.4	.3446	.3409	.3372	.3336	.3300	.3264	.3228	.3192	.3156	.3121
0.5	.3085	.3050	.3015	.2981	.2946	.2912	.2877	.2843	.2810	.2776
0.6	.2743	.2709	.2676	.2643	.2611	.2578	.2546	.2514	.2483	.2451
0.7	.2420	.2389	.2358	.2327	.2296	.2266	.2236	.2206	.2177	.2148
0.8	.2119	.2090	.2061	.2033	.2005	.1977	.1949	.1922	.1894	.1867
0.9	.1841	.1814	.1788	.1762	.1736	.1711	.1685	.1660	.1635	.1611
1.0	.1587	.1562	.1539	.1515	.1492	.1469	.1446	.1423	.1401	.1379
1.1	.1357	.1335	.1314	.1292	.1271	.1251	.1230	.1210	.1190	.1170
1.2	.1151	.1131	.1112	.1093	.1075	.1056	.1038	.1020	.1003	.0985
1.3	.0968	.0951	.0934	.0918	.0901	.0885	.0869	.0853	.0838	.0823
1.4	.0808	.0793	.0778	.0764	.0749	.0735	.0722	.0708	.0694	.0681
1.5	.0668	.0655	.0643	.0630	.0618	.0606	.0594	.0582	.0571	.0559
1.6	.0548	.0537	.0526	.0516	.0505	.0495	.0485	.0475	.0465	.0455
1.7	.0446	.0436	.0427	.0418	.0409	.0401	.0392	.0384	.0375	.0367
1.8	.0359	.0352	.0344	.0336	.0329	.0322	.0314	.0307	.0301	.0294
1.9	.0287	.0281	.0274	.0268	.0262	.0256	.0250	.0244	.0239	.0233
2.0	.0228	.0222	.0217	.0212	.0207	.0202	.0197	.0192	.0188	.0183
2.1	.0179	.0174	.0170	.0166	.0162	.0158	.0154	.0150	.0146	.0143
2.2	.0139	.0136	.0132	.0129	.0125	.0122	.0119	.0116	.0113	.0110
2.3	.0107	.0104	.0102	.0099	.0096	.0094	.0091	.0089	.0087	.0084
2.4	.0082	.0080	.0078	.0075	.0073	.0071	.0069	.0068	.0066	.0064
2.5	.0062	.0060	.0059	.0057	.0055	.0054	.0052	.0051	.0049	.0048
2.6	.0047	.0045	.0044	.0043	.0041	.0040	.0039	.0038	.0037	.0036
2.7	.0035	.0034	.0033	.0032	.0031	.0030	.0029	.0028	.0027	.0026
2.8	.0026	.0025	.0024	.0023	.0023	.0022	.0021	.0021	.0020	.0019
2.9	.0019	.0018	.0017	.0017	.0016	.0016	.0015	.0015	.0014	.0014
3.0	.00135									
3.5	.000233									
4.0	.0000317									
4.5	.00000340									
5.0	.000000287									

ANSWERS TO HOMEWORK EXERCISES

Chapter 1

1. Qualitative research deals with non-numerical information and is often typified by the case study (e.g., in-depth analysis of a group, situation, or event). Quantitative research deals with measurement on a numerical scale (how much, how many) and is exemplified by survey or experimental research.
2.
   a. part one -- survey; part two -- interview
   b. A survey is a quantitative method and part two is a qualitative method (an in-depth interview)
   c. Some advantages to the quantitative approach include its objectivity, replicability, time, and cost efficiency. Some advantages to the qualitative approach include in-depth understanding of the phenomenon under study and the generation of theories and hypotheses based on real-life observations rather than arm-chair speculations.
   d. The comparisons between the quantitative and qualitative methods is not one of superiority but of appropriateness. The best method to use depends on the question being asked and the stage of the research. If the researcher is in an exploratory stage, most likely qualitative techniques are more useful in gathering in-depth insight. In contrast, if the researcher is in a confirmatory stage, quantitative approaches provide a better route to gather objective evidence. Often researchers combine both methods in a way that makes the best sense for the study they want to conduct.
3.
   a. quantitative -- experiment
   b. qualitative -- in-depth analysis through interviews
   c. quantitative -- survey
   d. qualitative -- the researcher is gathering in-dept data through two case studies via participant observation
4. Inferential; descriptive
5.
   a. inferential -- inferring next year's prices from a sample semester
   b. descriptive -- describing the percentage of voters
   c. descriptive -- describing the batter's average number of hits
   d. inferential -- inferring the future probability of rape based on the previous ten years' data

6. a. No. Not all women in the United States are represented in the *Cosmopolitan* survey. In other words, there was a non-random sample used to generalize to all women in the United States. Only those women who read that issue of Cosmopolitan *and* chose to complete and return the survey are represented. In order to substantiate that all women in the United States felt happy or unhappy in their relationships every woman in the United States would have had to have been given an equal opportunity to participate in the survey, which they were not.
   b. The lawyers need to be cautious with how they apply the results from the experiment. The results most likely can only be generalized (applied) to university students (the target population of the experiment). In addition, the students evaluated "scenarios," not real-life cases. The results may vary across ethnic groups as well as according to the strength of ethnic identification of individuals within the groups. The investigation gives something for lawyers to think about but more evidence is needed before the lawyers should radically change their procedure.
   c. Again, the numbers are not representative of everyone in the United States. It is biased to those who have televisions, watch CNN, and are willing to call in and pay a minimal fee to cast their vote. In addition, the same person may have called in several times. In order to represent all people in the United States, CNN would need to gather a random sample (all people have an equal chance of being selected) of U.S. citizens.

Chapter 2

1. a. experiment
   b. surveys
   c. field research
   d. content analysis
2. population; sample
3. population -- all elementary school children; sample -- children in thirty classrooms
4. random or representative
5. a. non-random -- not all college student at this university (the population) had a chance of being included in the sample
   b. random -- all university professors had an equal chance of being included in the sample

ANSWERS TO HOMEWORK EXERCISES   301

      c.   random -- all *TV Guide* covers from the past ten years (the population) had an equal chance of being included in the sample

6. Random sampling refers to a way of selecting people (units) for a sample from a population such that each member of the population has an equal chance of being included in the population. In contrast, random assignment is a process of randomly putting people into conditions or groups in an experiment.

7. The independent variable (what the researcher varies) consists of the presence of speech training (present, absent) and the dependent variable (what the researcher measures) is the evaluation of the speeches.

8. The independent variable is the type of store (department, specialty) and the dependent variable is the amount of profit.

9. The independent variable consists of the presence of extra help (present, absent) and the dependent variable is midterm exam score.

10.   a.   interval
       b.   nominal
       c.   ordinal
       d.   ration

11. nominal (can put observations into categories only, but not tell if one is greater or less than the other), ordinal (can compare greater or less than among observations, but not how much greater or less than), interval (know how much greater or less than observations are compared to each other, but have an arbitrary bottom or zero point), ratio (know how much greater or less than observations are compared to each other and have an absolute fixed zero point)

12. The researcher can give you an exam on the material covered in this courses and if grades are assigned the measurement is ordinal and if scores are assigned the measurement is interval.

13.   a.   nominal
       b.   i.   ratio
            ii.   ordinal
       c.   i.   ratio
            ii.   ordinal
            iii.   nominal
       d.   nominal
       e.   ratio
       f.   nominal
       g.   i.   interval
            ii.   ordinal
            iii.   nominal
       h.   ratio
       i.   nominal
       j.   i.   ordinal

302 ANSWERS TO HOMEWORK EXERCISES

        ii.   ordinal
        iii.   interval
  k. interval

Chapter 3

1. A collection of measurements (data points) ordered numerically (from low to high or high to low)
2. score; frequency
3. frequency polygon -- line graph; histogram -- bar graph.
   With both types of graphs, the data points are plotted according to their frequency. With the frequency polygon, the plotted points are connected by a line, and with a histogram each plotted point has a bar drawn under it.
4. A normal distribution is perfectly symmetrical around the middle whereas a skewed distribution has one or the other tail pulled out.

5.

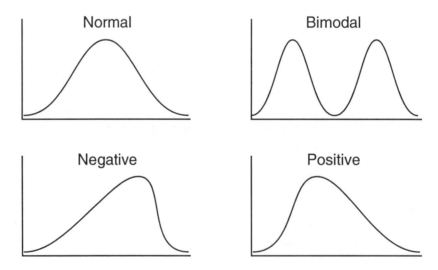

ANSWERS TO HOMEWORK EXERCISES

6. c -- scores on a very easy test

7. $\overline{X} = \dfrac{\sum X}{N} = \dfrac{(10 \times 35,000) + (15 \times 45,000) + 20 \times 60,000)}{45} = \$49,444$

8.
   a. median
   b. mean
   c. mode
   d. mode
   e. mean
   f. lower
   g. higher
   h. mean
   i. none, they all fall at exactly the same point -- the center
   j. mode

9.
   a. mode, median
   b. mode, median, mean
   c. mode only

10.
    a. flat curve (left)
    b. peaked curve (right)
    c. peaked curve (right)
    d. flat curve (left)

11.
    a. $V = SD^2$ or $SD = \sqrt{V} = \sqrt{225} = 15$
    b. $V = SD^2 = 11^2 = 121$
    c. b is the more homogeneous because the standard deviation and variance are smaller

12.
    a. variance
    b. median
    c. range

d. standard deviation
e. mean
f. mode
13. b, d, and e are distributions of numbers (because they are ordered from low to high or high to low)
14. a. $\bar{X} = \dfrac{\sum X}{N} = \dfrac{78 + 69 + 92 + 83 + 55 + 76}{6} = 75.5$

order numbers to form a distribution: 55, 69, 76, 78, 83, 92
middlemost score is between 76 and 78; Mdn = 77
Mo = none (no score occurs more frequently than any other score)
R = 92 - 55 = 37
$\bar{X}$ = 75.5; Mdn = 77; Mo = none; R = 37
b. $\bar{X}$ = 76.6; Mdn = 79; Mo = none; R = 25
c. $\bar{X}$ = 84.3; Mdn = 83; Mo = 82 and 94; R = 24
d. $\bar{X}$ = 76.8; Mdn = 73; Mo = 65; R = 34
e. $\bar{X}$ = 81.50; Mdn = 81; Mo = 81; R = 14
f. $\bar{X}$ = 72; Mdn = 67; Mo = 67; R = 15
g. $\bar{X}$ = 3.17; Mdn = 3; Mo = 3; R = 5

Chapter 4

1. sample distribution
   a. $\bar{X}$
   b. SD
   c. $z = \dfrac{X - \bar{X}}{SD}$
2. population distribution
   a. μ
   b. σ
   c. $z = \dfrac{X - \mu}{\sigma}$
3. sampling distribution
   a. μ
   b. $\sigma_{\bar{x}}$

ANSWERS TO HOMEWORK EXERCISES **305**

    c. $z = \dfrac{\overline{X} - \mu}{\sigma_{\overline{x}}}$

4.   distribution of differences
    a. μ
    b. σ_{diff}
    c. $z = \dfrac{(\overline{X}_1 - \overline{X}_2) - 0}{\sigma_{diff}}$

5.   sample; population
6.   use statistics -- sample distribution only;
    use parameters -- population distribution, sampling distribution, distribution of differences
    The distributions that use statistics are based on observations from a sample, whereas distributions that use parameters are based on all observations from a population.
7.   mean
8.   standard deviation
9.   the standard error of the mean
10.   the standard error of the difference (or the standard error of the mean difference)
11.   a.   <u>difference score</u>
              11 - 3 =   +8
              8 - 12 =   -4
              16 - 12 =   +4
              9 - 13 =   -4
              8 - 16 =   -8
              5 - 5 =   0
              11 - 11 =   0
              14 - 10 =   +4

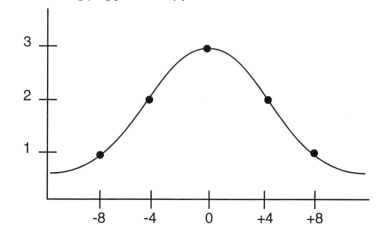

306 ANSWERS TO HOMEWORK EXERCISES

	b.	distribution of differences
	c.	$\mu = 0$
	d.	standard error of the (mean) difference
12.		standard deviation
13.	a.	$z = +1.50$
	b	$z = -2.00$
	c.	$z = 0.00$
	d.	$z = -0.66$
	e.	$z = +2.75$

14.

a. R=18; 6–24

b. R=1500; 250–1750

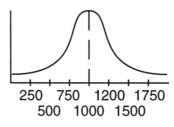

c. R=30; 60–90

d. R=90; 105-195

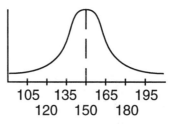

15.   a.

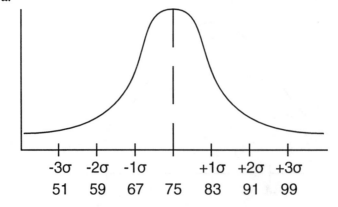

ANSWERS TO HOMEWORK EXERCISES **307**

b.   i. $z = \dfrac{X - \overline{X}}{SD} = \dfrac{83 - 75}{8} = \dfrac{8}{8} = +1.00$
     ii.   z = -1.50
     iii.   z = +2.25
     iv.   z = 0.00
     v.   z = -1.88
c.   i.   68%
     ii.   99%
     iii.   50%
     iv.   16%
     v.   16%
     vi.   97.5%
     vii.   95%
     viii. 99.5%
d.   i.   87
     ii.   75
     iii.   51
     iv.   73
     v.   93.64
     vi.   61

16. a.

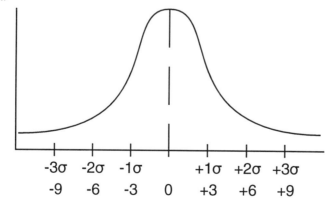

     b.   3

        c.   -6
17.   a.   the NFL
        b. μ, σ
        c.   232.5
        d.   sample distribution
        e. $\overline{X}$, SD
        f.   0.5
        g.   68%
        h.   2.5%
        i.   28; note that actually the sampling distribution is created with an infinite number of samples
        j.   sampling distribution
        k. 3σ
        l.   0.5%

Chapter 5

1. the null hypothesis says there is not a difference among observed phenomena (or that differences are due to chance) while the alternative (research) hypothesis says there is a difference.
2. the alternative hypothesis
3. the null hypothesis
4.   a.   the null hypothesis
      b.   the alternative hypothesis
5.   a. H_a; H_0; one-tail test
      b. H_a; H_0; two-tail test
      c. H_a; H_0; one-tail test
6.   a. H_a
      b. H_0
      c. H_a
      d. H_a
      e. H_0
7.   a. $H_a\ \mu_{stu} \neq \mu_{prof}$; $H_0: \mu_{stu} = \mu_{prof}$
      b. $H_a\ \mu_{pov} > \mu_{nonp}$; $H_0: \mu_{pov} \leq \mu_{nonp}$
      c. $H_a\ \mu_{men} \neq \mu_{wom}$; $H_0: \mu_{men} = \mu_{wom}$
      d. $H_a\ \mu_{pri} \neq \mu_{pub}$; $H_0: \mu_{pri} = \mu_{pub}$
8.   a.   snow, $p = .01$
      b.   overcast, $p = .25$
      c.   fog, $p = .50$
9.   a.   French Admiral, $p = .51$
      b.   Sapphire, $p = .70$
      c.   Majestic Prince, $p = .99$
10. $p < .05$; $p < .01$;
The probabilities levels refer to the researcher's willingness to accept that 1 or 5 times out of 100 their conclusion regarding the null hypothesis could be wrong or that there is a 95% or 99% chance that the alternative hypothesis is correct. At these established levels the researcher is able to reject the null hypothesis.
11. two-tail test
12. one-tail test
13.   a.   one-tail test
       b.   two-tail test
       c.   one-tail test
       d.   two-tail test
14.   a. ± 1.96
       b. ± 2.58
       c. $+1.65$ or -1.65
       d. $+2.33$ or -2.33

15. reject H_0
16.
   a. do not reject (accept) H_0
   b. reject H_0
   c. do not reject (accept) H_0
   d. reject H_0
   e. do not reject (accept) H_0
   f. do not reject (accept) H_0
   g. reject H_0
17.
   a. reject H_0
   b. do not reject (accept) H_0
   c. do not reject (accept) H_0
   d. reject H_0
   e. do not reject (accept) H_0
   f. do not reject (accept) H_0
   g. reject H_0
18. Type I error
19. Type II error
20. .05; .01
21. more
22. $z = \dfrac{\overline{X}_1 - \overline{X}_2}{\sigma_{diff}} = \dfrac{135 - 115}{5} = \dfrac{20}{5} = 4.00$

Chapter 6

1. $t = \dfrac{\overline{X}_1 - \overline{X}_2}{\sigma_{diff}}$
2. $df = (N_1 - 1) + (N_2 - 1)$ or $df = N_1 + N_2 - 2$
3. distribution of differences
4. degrees of freedom, one- or two-tail test, significance level
5. rejects
6. significantly
7.
   a. $df = 8$; do not reject H_0 with critical value of $t_{(8).05} = 2.306$
   b. $df = 20$; do not reject H_0 with critical value of $t_{(20).05} = 2.086$
   c. $df = 28$; reject H_0 at $p < .05$ with critical value of $t_{(28).05} = 2.048$
   d. $df = 30$; reject H_0 at $p < .05$ with critical value of $t_{(30).05} = 2.042$
8. the larger the sample size (N) the greater your ability to reject H_0; a statistical test is more powerful with larger sample sizes

9.
   a. $N = 24$
   b. $N = 61$
   c. $N = 15$
   d. $N = 8$
   e. $N = 12$
   f. $N = 21$

10.
   a. two-tail hypothesis; the direction of the means is not specified
   b. $H_a: \mu_1 \neq \mu_2;\ H_0: \mu_1 = \mu_2$
   c. $t = \dfrac{\overline{X}_1 - \overline{X}_2}{\sigma_{diff}} = \dfrac{4 - 6.5}{1.20} = \dfrac{-2.5}{1.20} = -2.08$
      $df = (N_1 - 1) + (N_2 - 1) = (10 - 1) + (10 - 1) = 9 + 9 = 18$
   d. $t_{(18).05} = 2.101;\ t_{(18).01} = 2.878$
   e. Since the calculated value $t = -2.08$ is not larger (in absolute terms) than the critical value at the $p < .05$ level, do not reject H_0; people do not react differently to a negative political ad than to a neutral political ad.
   f. You would recommend either type of ad based on this data; it makes no difference.
   g. $H_a: \mu_1 < \mu_2;\ H_0: \mu_1 \geq \mu_2$
   h. Since the calculated value $t = -2.08$ is larger (in the predicted direction) than the critical value for the one-tail test $t_{(18).05} = 1.734$, reject H_0 at the $p < .05$ level (note: you cannot reject at the preferred $p < .01$ level with a critical value $t_{(18).01} = 2.552$).

11.
   a. two-tail hypothesis
   b. $H_a: \mu_A \neq \mu_B;\ H_0: \mu_A = \mu_B$
   c. $t = \dfrac{\overline{X}_1 - \overline{X}_2}{\sigma_{diff}} = \dfrac{42.60 - 38.62}{1.36} = \dfrac{-3.98}{1.36} = 2.93$
      $df = (N_1 - 1) + (N_2 - 1) = (31 - 1) + (31 - 1) = 30 + 30 = 60$
      critical values: $t_{(60).05} = 2.00;\ t_{(60).01} = 2.66$
      Since the calculated value of $t = 2.93$ is larger (in absolute terms) than the critical value $t_{(60).05} = 2.00$, reject H_0 at $p < .05$. Can we do any better? Yes, since the calculated value $t = 2.93$ is larger than the critical value at the $p < .01$ level, $t_{(60).01} = 2.66$, reject H_0 at $p < .01$.
   d. reject H_0 at $p < .01$; there is a difference in attitudes directed toward a minority group between those who watched the film and those who did not.

12.
   a. two-tail hypothesis; the direction of the means is not specified
   b. $H_a: \mu_1 \neq \mu_2;\ H_0: \mu_1 = \mu_2$
   c. $t = \dfrac{\overline{X}_1 - \overline{X}_2}{\sigma_{diff}} = \dfrac{98.06 - 102.35}{2.45} = \dfrac{-4.29}{2.45} = -1.75$

$df = (N_1 - 1) + (N_2 - 1) = (61 - 1) + (61 - 1) = 60 + 60 = 120$
critical values: $t_{(120).05} = 1.98$; $t_{(120).01} = 2.517$
Since the calculated value of $t = -1.75$ is not larger (in absolute terms; ignore the sign) than the critical value $t_{(120).05} = 1.98$, do not reject H_0.

  d.   do not reject H_0; the individual therapy program did not significantly affect delinquent boys' levels of measured anxiety.

13.   a.   one-tail hypothesis; the direction of the means is predicted (that the *Brady Bunch* group will have a higher mean than the *Three's Company* group)

  b. H_a: $\mu_{BB} > \mu_{TC}$; H_0: $\mu_{BB} \leq \mu_{TC}$

  c. $t = \dfrac{\overline{X}_1 - \overline{X}_2}{\sigma_{diff}} = \dfrac{3.9 - 1.8}{0.82} = \dfrac{2.1}{0.82} = 2.56$

$df = (N_1 - 1) + (N_2 - 1) = (31 - 1) + (31 - 1) = 30 + 30 = 60$
critical values: $t_{(60).05} = 1.671$; $t_{(60).01} = 2.390$
Since the calculated value $t = 2.56$ is larger (in the predicted direction) than the critical value $t_{(60).05} = 1.671$, reject H_0 at $p < .05$. Can we do any better? Yes. Since the calculated value $t = 2.56$ is larger than the critical value at the .01 level $t_{(60).01} = 2.390$, reject H_0 at $p < .01$.

  d.   reject H_0 at $p < .01$; watching the *Brady Bunch* causes people to want to have significantly more children than does watching *Three's Company*

14.   a.   one-tail hypothesis; the direction of the means is predicted (that the mean number of seconds before initating a conversation will be lower with the female than with the male)

  b. H_a: $\mu_F < \mu_M$; H_0: $\mu_F \geq \mu_M$

  c.   No. The standard error of the difference is negative (which can never happen). Also, if you use our preferred second method for conducting a t-test you should notice that the means are not in the predicted direction; therefore, you stop right there before ever calculating a value for t.

  d.   The means are in the predicted direction (the female has a lower mean than the male).

$t = \dfrac{\overline{X}_1 - \overline{X}_2}{\sigma_{diff}} = \dfrac{5 - 9}{2.30} = \dfrac{-4}{2.30} = -1.74$

$df = (N_1 - 1) + (N_2 - 1) = (13 - 1) + (13 - 1) = 12 + 12 = 24$
critical values: $t_{(24).05} = 1.711$; $t_{(24).01} = 2.492$
Since the calculated value $t = -1.74$ is larger (in the predicted direction) than the critical value $t_{(24).05} = 1.711$, reject H_0 at $p < .05$. Can we do any better? No. Since the calculated value $t =$

-1.74 is not larger than the critical value at the .01 level $t_{(24).01} = 2.492$.

e. reject H_0 at $p < .05$; men initiate a conversation with a female significantly faster than with a male

f.
Source	SS	df	MS	F
Between	25	3	8.33	3.53
Within	85	36	2.36	
Total	110	39		

degrees of freedom are same as part c above

$$MS_B = \frac{SS_B}{df_B} = \frac{25}{3} = 8.33$$

$$MS_W = \frac{SS_W}{df_W} = \frac{85}{36} = 2.36$$

$$F = \frac{MS_B}{MS_W} = \frac{8.33}{2.36} = 3.53$$

critical values: $F_{(3,36).05} = 2.86$; $F_{(3,36).01} = 4.38$
Since the calculated value $F = 3.53$ is larger than the critical value $F_{(3,36).05} = 2.86$, reject H_0 at $p < .05$. Can we do any better? No. The calculated value $F = 3.53$ is not larger than the critical value at the .01 level $F_{(3,36).01} = 4.38$.

g. reject H_0 at $p < .05$; there is some difference in beliefs about life on other planets among children who watch different *Star Trek* programs.

h. cannot be determined without conducting a post-hoc test

15. a. H_a: $\mu_5 \neq \mu_{10} \neq \mu_{15}$
H_0: $\mu_5 = \mu_{10} = \mu_{15}$

b. $N = 30$ since there were 10 employees per group.

c.
Source	SS	df	MS	F
Between	180	2	90	3.04
Within	800	27	29.63	
Total	980	29		

$df_B = K - 1 = 3 - 1 = 2$
$df_W = N - K = 30 - 3 = 27$
$df_T = N - 1 = 30 - 1 = 29$ (or $df_T = df_B + df_W = 2 + 27 = 29$)

$$MS_B = \frac{SS_B}{df_B} = \frac{180}{2} = 90$$

$$MS_W = \frac{SS_W}{df_W} = \frac{800}{27} = 29.63$$

$$F = \frac{MS_B}{MS_W} = \frac{90}{29.63} = 3.04$$

critical values: $F_{(2,27).05} = 3.35$; $F_{(2,27).01} = 5.49$
Since the calculated value $F = 3.04$ is not larger than the critical value $F_{(2,27).05} = 3.35$, do not reject H_0.

d. do not reject H_0; there is no difference in productivity rates as a function of positive reinforcement from supervisors

e. it doesn't matter since there was no difference among the three groups

f.
Source	SS	df	MS	F
Between	180	2	90	6.41
Within	800	57	14.04	
Total	980	59		

$df_B = K - 1 = 3 - 1 = 2$
$N = 20 \times 3 = 60$ employees in the study
$df_W = N - K = 60 - 3 = 57$
$df_T = N - 1 = 60 - 1 = 59$ (or $df_T = df_B + df_W = 2 + 57 = 59$)

$$MS_B = \frac{SS_B}{df_B} = \frac{180}{2} = 90$$

$$MS_W = \frac{SS_W}{df_W} = \frac{800}{57} = 14.04$$

$$F = \frac{MS_B}{MS_W} = \frac{90}{14.04} = 6.41$$

critical values: $F_{(2,57).05} \approx 3.16$; $F_{(2,57).01} \approx 5.00$
Since the calculated value $F = 6.41$ is larger than the critical value $F_{(2,57).05} = 3.16$, reject H_0 at $p < .05$. Can we do any better? Yes. Since the calculated value $F = 6.41$ is larger than the critical value at the .01 level $F_{(2,57).01} \approx 5.00$, reject H_0 at $p < .01$.

g. reject H_0 at $p < .01$; there is some difference in workers' productivity rates as a function of time receiving positive praise from supervisors
h. cannot be determined without conducting a post-hoc test

Chapter 8

1. multiple-factor analysis of variance (F-test)
2.
   a. 2
   b. 5
   c. 2
   d. 3
   e. 3
   f. 4
3.
   a. 2
   b. 9
   c. 3
   d. 3
   e. 0, there is no fourth factor
   f. 2
4.
   a. 4 x 3 = 12
   b. 2 x 2 x 2 = 8
   c. 2 x 3 x 4 = 24
5.
   a. 3 x 2 x 3 = 18 conditions x 5 subjects per condition = 90 total subjects
   b. 4 x 3 = 12 conditions x 16 subjects per condition = 192 total subjects
   c. 3 x 3 x 3 = 27 conditions x 20 subjects per condition = 540 total subjects
6. main effect
7. interaction effect
8.
   a. There are three independent variables each with two levels: type of conference (face-to-face, televised), type of firm (national, international), type of employee (management, regular), so this is a 2 x 2 x 2 factorial design.
   b. Since there are three independent variables there are three possible main effects.
   c. There are a total of four possible interaction effects: three two-way interactions (conference x firm, conference x employee, firm x employee) and one three-way interaction (conference x firm x employee).

9. a. three independent variables: grade level has three levels (kindergarten, 3rd grade, 5th grade), number of acts of violence viewed has three levels (0, 5, 10), and post-viewing activity has four levels (read, play, watch TV, sit)
   b. one dependent variable: number of hits and kicks (ratio-level data)
   c. 3 x 3 x 4 factorial design (see part a above)
   d. multiple-factor analysis of variance
   e. Since there were a total of 180 children in the study who were divided equally (randomly assigned) to the three conditions of the independent variable number of violent acts viewed, there were $\frac{180}{3}$ = 60 children who watched 5 acts of violence.
   f. This answer may be easier to understand visually by examining the number of subjects per condition:

Kindergarten	0 acts	5 acts	10 acts
read	N=5	N=5	N=5
play	N=5	N=5	N=5
TV	N=5	N=5	N=5
sit	N=5	N=5	N=5

3rd grade	0 acts	5 acts	10 acts
read	N=5	N=5	N=5
play	N=5	N=5	N=5
TV	N=5	N=5	N=5
sit	N=5	N=5	N=5

5th grade	0 acts	5 acts	10 acts
read	N=5	N=5	N=5
play	N=5	N=5	N=5
TV	N=5	N=5	N=5
sit	N=5	N=5	N=5

   A total of N = 15 5th graders sat in a room.
   g. interaction effect because more than one factor (independent variable) is involved: number of acts of violence viewed and post-viewing activity.
   h. main effect because there is only one factor involved: grade level

10. a. main effect for marital status:
   H_a: $\mu_{marr} \neq \mu_{sing}$
   H_0: $\mu_{marr} = \mu_{sing}$
   main effect for program type:
   H_a: $\mu_{marr} \neq \mu_{div}$
   H_0: $\mu_{marr} = \mu_{div}$
   interaction effect for marital status x program type:

b. H_a: $\mu_{marr\text{-}marr} - \mu_{marr\text{-}div} \neq \mu_{sing\text{-}marr} - \mu_{sing\text{-}div}$

H_0: $\mu_{marr\text{-}marr} - \mu_{marr\text{-}div} = \mu_{sing\text{-}marr} - \mu_{sing\text{-}div}$

the distribution of subjects to conditions is as follows:

Marital Status

	Married	Single	
Marriage	N=12	N=18	N=30
Divorce	N=12	N=18	N=30
	N=24	N=36	N=60

Type of Program

So, N = 12 married men saw the program about marriage.

c.
Source	SS	df	MS	F
Marital Status	8.2	1	8.2	4.1
Program Type	7.5	1	7.5	3.75
Marital x Program	14.7	1	14.7	7.35
Error	112	56	2	
Total	142.4	59		

Since the total sum of squares (SS_T) is the sum of all the between-groups sum of squares and the error sum of squares, SS_{Error} can be computed as follows:

SS_{Error} = 142.4 - (8.2 + 7.5 + 14.7) = 142.4 - 30.4 = = 112

$df_{mar} = K_{mar} - 1 = 2 - 1 = 1$
$df_{prog} = K_{prog} - 1 = 2 - 1 = 1$
$df_{marxprog} = (K_{mar} - 1) \times (K_{prog} - 1) = (2 - 1) \times (2-1) = 1 \times 1 = 1$
$df_{Error} = N - K_{mar} K_{prog} = 60 - (2)(2) = 60 - 4 = 56$
$df_T = N - 1 = 60 - 1 = 59$

$$MS_{mar} = \frac{SS_{mar}}{df_{mar}} = \frac{8.2}{1} = 8.2$$

$$MS_{prog} = \frac{SS_{prog}}{df_{prog}} = \frac{7.5}{1} = 7.5$$

$$MS_{marxprog} = \frac{SS_{marxprog}}{df_{marxprog}} = \frac{14.7}{1} = 14.7$$

$$MS_{Error} = \frac{SS_{Error}}{df_{Error}} = \frac{112}{56} = 2$$

$$F_{mar} = \frac{MS_{mar}}{MS_{mar}} = \frac{8.2}{2} = 4.1$$

$$F_{prog} = \frac{MS_{prog}}{MS_{prog}} = \frac{7.5}{2} = 3.75$$

$$F_{marxprog} = \frac{MS_{marxprog}}{MS_{marxprog}} = \frac{14.7}{2} = 7.35$$

critical values for all three statistical tests are the same because the degrees of freedom are the same (1 and 56): $F_{(1,56).05} \approx 4.02$; $F_{(1,56).01} \approx 7.12$

Test of the main effect of marital status:
calculated value $F = 4.1$ is larger than critical value $F_{(1,56).05} \approx 4.02$, so reject H_0 at $p < .05$ (cannot do better at $p < .01$)

Test of the main effect of program type:
calculated value $F = 3.75$ is not larger than critical value $F_{(1,56).05} \approx 4.02$, do not reject H_0

Test of the main effect of marital status:
calculated value $F = 7.35$ is larger than critical value $F_{(1,56).05} \approx 4.02$, so reject H_0 at $p < .05$; can do better: calculated value $F = 7.35$ is larger than critical value at .01 level $F_{(1,56).01} \approx 7.12$, so reject H_0 at $p < .01$)

d. yes -- main effect for marital status and interaction effect
e. main effect for marital status: reject H_0 at $p < .05$
main effect for program type: do not reject H_0
interaction effect: reject H_0 at $p < .01$
main effect for marital status: depression differs among men as a function of marital status (married vs. single), independent of program viewed
interaction effect: men differ significantly in their depression level as a function of the combination of their marital status and the type of program viewed

f.

Source	SS	df	MS	F
Marital Status	10.8	1	10.8	3.6
Program Type	12.6	1	12.6	4.2
Marital x Program	9.8	1	9.8	3.27
Error	20.4	68	3	
Total	237.2	71		

critical values: $F_{(1,68).05} \approx 3.98$; $F_{(1,68).01} \approx 7.02$

main effect for marital status: do not reject H_0
main effect for program type: reject H_0 at $p < .05$
interaction effect: do not reject H_0

g. yes -- main effect for program type only
h. see part f. above
main effect for program type: depression differed significantly among men who viewed the program about divorce versus those who viewed the program about marriage, independent of marital status

11. a. there are three independent variables: gender (male, female), exposure (5 minutes, 10 minutes, 15 minutes), and presence of extra warning (present, absent), so this is a 2 x 3 x 2 factorial design
b. three possible main effects since there are three independent variables
c. there are four possible interaction effects
   three two-way interactions: gender x exposure, gender x warning, exposure x warning
   one three-way interaction: gender x exposure x warning

d.

Source	SS	df	MS	F
Gender	40	1	40	6.67
Exposure	30	2	15	2.50
Warning	20	1	20	3.33
Gend. x Expos.	60	2	30	5.00
Gend. x Warn.	20	1	20	3.33
Expos. x Warn.	20	2	10	1.67
Ge. x Ex. x Wa.	80	2	40	6.67
Error	648	108	6	
Total	918	119		

critical values for assessing effects with 1 and 108 degrees of freedom: $F_{(1,108).05} \approx 3.94$; $F_{(1,108).01} \approx 6.90$
critical values for assessing effects with 2 and 108 degrees of freedom: $F_{(2,108).05} \approx 3.09$; $F_{(2,108).01} \approx 4.82$

Effect	Calculated F	Critical F $p<.05$	Critical F $p<.01$	decision
Gender	6.67	3.94	6.90	reject H_0, $p<.05$
Exposure	2.50	3.09	4.82	do not reject H_0
Warning	3.33	3.94	6.90	do not reject H_0
Gend. x Expos.	5.00	3.09	4.82	reject H_0, $p<.01$
Gend. x Warn.	3.33	3.94	6.90	do not reject H_0
Expos. x Warn.	1.67	3.09	4.82	do not reject H_0
Ge. x Ex. x Wa.	6.67	3.09	4.82	reject H_0, $p<.01$

e. yes, main effect for gender, interaction effect between gender and exposure, interaction effect between gender, exposure, and warning

f. see chart above, part d

main effect for gender: males and females exert significantly different levels of self control

interaction of gender x exposure: ability to exert self control is based on a combined effect between gender and exposure

interaction of gender x exposure x warning: ability to exert self control is based on a combined effect among all three variables: gender, exposure, warning

Chapter 9

1. simple correlation
2. a. -1.00 to +1.00
   b. direction
   c. increases
   d. decreases
   e. the strength of the association
3. a. no correlation
   b. perfect, negative
   c. moderate, positive
   d. perfect, positive
   e. low, negative
4. Pearson product-moment correlation coefficient; r
5. Spearman's rho
6. a. $r = 0.78$
   b. $r = -0.13$
   c. $r = -1.00$
   d. $r = -0.92$
   e. $r = -0.21$

	f.	equal
7.	a.	$H_0: \rho = 0$; $H_a: \rho \neq 0$
	b.	df = N - 2 = 30 - 2 = 28

7.
  a. $H_0: \rho = 0$; $H_a: \rho \neq 0$
  b. df = N - 2 = 30 - 2 = 28
     critical values: $r_{(28).05} = .361$; $r_{(28).01} = .463$
     Since the calculated value r = -.43 is larger (in absolute terms) than the critical value at the .05 level (but not the .01 level) $r_{(28).05} = .361$, reject H_0 at $p < .05$
  c. moderate, negative correlation
  d. the more time computer consultants spend on the computer the worse their vision tends to be

8.
  a. $H_0: \rho = 0$; $H_a: \rho \neq 0$
  b. df = N - 2 = 52 - 2 = 50
     critical values: $r_{(50).05} = .250$; $r_{(50).01} = .325$
     Since the calculated value r = .24 is not larger than the critical value $r_{(50).05} = .250$, do not reject (accept) H_0
  c. low, positive correlation
  d. there is no significant relationship between perceived intelligence and excessive talk

9.
  a. $H_0: \rho = 0$; $H_a: \rho \neq 0$
  b. df = N - 2 = 10 - 2 = 8
     critical values: $r_{(8).05} = .632$; $r_{(8).01} = .765$
     Since the calculated value r = .78 is larger than the critical value at the .01 level $r_{(8).01} = .765$, reject H_0 at $p < .01$
  c. high, positive correlation
  d. the more rainfall per year the more absenteeism there is from work

10. coefficient of determination; r^2

11. shaded area represents shared variance
  a. $r^2 = (.15)^2 = .02$
  b. $r^2 = (.6)^2 = .36$

12.
  a. high, positive; 71% [$r^2 = (.84)^2 = .706 = 71\%$]
  b. low, negative; 2%
  c. perfect, negative; 100%
  d. low, positive; 5%
  e. no correlation; 0%
  f. high, positive; 85%
  g. moderate, negative; 32%

13. multiple correlation

14. $R_{X \cdot YZ}$; $R^2_{X \cdot YZ}$; the percentage of variance accounted for in X by both Y and Z

15. partial correlation

16. $r_{UV \cdot Y}$; $r^2_{UV \cdot Y}$; the relationship between U and V while removing the effects of Y
17.
    a. N + O
    b. M + O + N
    c. N
    d. M + O + P
    e. O + P
    f. P
    g. N + O + P
    h. M
    i. M + O

Chapter 10

1. To predict one variable (Y) from another variable (X)
2. the regression line or line of best fit
3. Y = bX + a; Y = predicted score, b = slope of regression line, X = specific score, a = y-intercept
4. b (beta value); refers to the slope of the line
5. a (y-intercept); refers to where the line crosses the y-axis
6.
    a. 11.52
    b. 261
    c. 53.2
    d. 0.88
7. accuracy of the predicted value
8. negative -- as the correlation increases, error in prediction decreases, and as correlation decreases, error in prediction increases
9. small
10.
    a. r = .34
    b. r = -1.00
    c. r = .67
    d. r = -.49
    e. r = -.98
11.
    a. df = N - 2 = 30 - 2 = 28
       critical values: $r_{(28).05}$ = .361; $r_{(28).01}$ = .463
       Since the calculated value r = -0.28 is not larger (in absolute terms) than the critical value $r_{(28).05}$ = .361, do not reject H_0. The correlation is not significant so the researcher cannot predict sleep based on number of cigarettes smoked better than chance.
    b. df = N - 2 = 22 - 2 = 20
       critical values: $r_{(20).05}$ = .423; $r_{(20).01}$ = .537

Since the calculated value r = 0.47 is larger than the critical value $r_{(20).05}$ = .423, reject H_0 at $p < .05$ (but cannot reject at .01). The correlation is significant so the manager can predict employees' gossiping based on number of cups of coffee drank better than chance.

Chapter 11

1. parametric; nonparametric
2.
   a. interval, ratio
   b. nominal, ordinal, interval, ratio
   c. interval, ratio
   d. nominal
   e. interval, ratio
   f. ordinal, interval, ratio
   g. interval, ratio
   h. interval, ratio
   i. interval, ratio
   j. nominal
   k. ordinal
3. multiple-sample chi-square analysis: the sample variable has three levels for when first crime was committed (before age 18, age 18-30, over age 30), and the categorical variable, commission of subsequent crimes, has two levels (committed additional crimes, did not commit additional crimes)
4. one-sample (one-way design) assesses differences among categories
   multiple-sample (two-way design) assesses differences among samples
5.
   a. $\chi^2 = \Sigma \frac{(O-T)^2}{T}$
   b. $T = \frac{N}{K}$ where N = the total number of observations, and K = the total number of categories
   c. df = (C - 1) where C = number of columns, or, df = (R - 1) where R = number of rows
   d. one
   e. difference among categories
6.
   a. $\chi^2 = \Sigma \frac{(O-T)^2}{T}$
   b. $T = \frac{RxC}{G}$

324 ANSWERS TO HOMEWORK EXERCISES

    c.   df = (R - 1) x (C - 1)
    d.   two
    e.   difference among samples

7.   a.   multiple-sample χ^2
there are two variables: the sampling variable, class level, with four levels (freshman, sophomore, junior, senior), and the categorical variable, major, with 14 levels (economics, mathematics, communication, physics, psychology, sociology, English, French, German, art, music, law and society, chemistry, biology)
df = (R - 1) x (C - 1) = (4 - 1) x (14 - 1) = 3 x 13 = 39 df

    b.   log-linear analysis
there are three variables: gender of CEO (male, female), weekly time spent at work (< 40 hours, 40-50 hours, > 50 hours), and yearly income (< $50K, $50-100K, > $100K)

    c.   one-sample χ^2
one variable: letter grade in class with five levels
df = (C - 1) = (5 - 1) = 4

    d.   log-linear analysis
three variables: amusement park, primary ride, gross annual profit

    e.   one-sample χ^2
one variable: method of beginning a conversation with seven levels
df = (C - 1) = (7 - 1) = 6

    f.   multiple-sample χ^2;
two variables: gender with two levels and opinion of the death penalty with three levels
df = (R - 1) x (C - 1) = (2 - 1) x (3 - 1) = 1 x 2 = 2

8.   a.   df = (R - 1) x (C - 1) = (4 - 1) x (4 - 1) = 3 x 3 = 9
critical values: $\chi^2_{(9),.05} = 16.92$; $\chi^2_{(9),.01} = 21.67$
Since the calculated value $\chi^2 = 17$ is larger than the critical value $\chi^2_{(9),.05} = 16.92$ reject H_0 at $p < .05$. Can we do any better? No, since the calculate value $\chi^2 = 17$ is not larger than the critical value at the .01 level $\chi^2_{(9),.01} = 21.67$.
final conclusion: reject H_0 at $p < .05$

    b.   df = (C - 1) = (6 - 1) = 5
critical values: $\chi^2_{(5),.05} = 11.07$; $\chi^2_{(5),.01} = 15.09$
do not reject H_0

    c.   df = (3 - 1) x (8 - 1) = 2 x 7 = 14

critical values: $\chi^2_{(14).05} = 23.68$; $\chi^2_{(14).01} = 29.14$
reject H_0 at $p < .01$

d. df = (2 - 1) x (2 - 1) = 1 x 1 = 1
critical values: $\chi^2_{(1).05} = 3.84$; $\chi^2_{(1).01} = 6.63$
reject H_0 at $p < .05$

e. df = (3 - 1) = 2
critical values: $\chi^2_{(2).05} = 5.99$; $\chi^2_{(2).01} = 9.21$
reject H_0 at $p < .01$

f. df = (6 - 1) x (6 - 1) = 5 x 5 = 25
critical values: $\chi^2_{(25).05} = 37.65$; $\chi^2_{(25).01} = 44.31$
do not reject H_0

g. df = (10 - 1) = 9
critical values: $\chi^2_{(9).05} = 16.92$; $\chi^2_{(9).01} = 21.67$
reject H_0 at $p < .01$

9. one-sample χ^2
total number of employees = 19 + 15 + 10 + 9 + 9 = 62

theoretical (expected) value = $T = \dfrac{N}{K} = \dfrac{62}{5} = 12.4$

$$\chi^2 = \Sigma \dfrac{(O-T)^2}{T}$$

$$= \dfrac{(19-12.4)^2}{12.4} + \dfrac{(15-12.4)^2}{12.4} + \dfrac{(10-12.4)^2}{12.4} + \dfrac{(9-12.4)^2}{12.4} + \dfrac{(9-12.4)^2}{12.4}$$

$= 3.51 + 0.55 + 0.46 + 0.93 + 0.93$

$= 6.38$

df = (C - 1) = (5 - 1) = 4
critical values: $\chi^2_{(4).05} = 9.49$; $\chi^2_{(4).01} = 13.28$
Since the calculated value $\chi^2 = 6.38$ is not larger than the critical value
$\chi^2_{(4).05} = 9.49$, do not reject H_0
The employees did not have a preference for certain Broadway shows, and any observed differences were due to chance.

10. one-sample χ^2
a total of N = 40 + 32 + 18 + 23 = 113 students were surveyed

theoretical (expected) value = $T = \dfrac{N}{K} = \dfrac{113}{4} = 28.25$

$$\chi^2 = \Sigma \frac{(O-T)^2}{T}$$

$$= \frac{(40-28.25)^2}{28.25} + \frac{(32-28.25)^2}{28.25} + \frac{(18-28.25)^2}{28.25} + \frac{(23-28.25)^2}{28.25}$$

$$= 4.89 + 0.50 + 3.72 + 0.98$$

$$= 10.09$$

df = (C - 1) = (4 - 1) = 3

critical values: $\chi^2_{(3).05} = 7.81$; $\chi^2_{(3).01} = 11.34$

Since the calculated value $\chi^2 = 10.09$ is not than the critical value

$\chi^2_{(4).05} = 9.49$, do not reject H_0

There is a significant difference in students' selection of a geographical location.

11. multiple-sample χ^2

create the marginal frequencies using the observed frequencies:

	0-10 words	11-20 words	21-30 words	
No distractions				30
Television				30
TV + Talking				30
	15	20	55	90

these marginal frequencies must be maintained when creating the theoretical frequencies, using the formula $T = \frac{RxC}{G}$

for the 0-10 words, no distractions group:

$$T = \frac{30 \times 15}{9} = \frac{450}{9} = 5$$

in this way we can obtain the theoretical frequencies for each of the nine cells in the table as follows:

	0-10 words	11-20 words	21-30 words	
No distractions	5	6.67	18.33	30
Television	5	6.67	18.33	30
TV + Talking	5	6.67	18.33	30
	15	20	55	90

ANSWERS TO HOMEWORK EXERCISES **327**

$$\chi^2 = \Sigma \frac{(O-T)^2}{T}$$

$$= \frac{(0-5)^2}{5} + \frac{(5-5)^2}{5} + \frac{(10-5)^2}{5} + \frac{(5-6.67)^2}{6.67} + \frac{(5-6.67)^2}{6.67}$$

$$+ \frac{(10-6.67)^2}{6.67} + \frac{(25-18.33)^2}{18.33} + \frac{(20-18.33)^2}{18.33} + \frac{(10-18.33)^2}{18.33}$$

$$= 5 + 0 + 5 + 0.42 + 0.42 + 1.66 + 2.43 + 0.15 + 3.79$$

$$= 18.87$$

df = (R - 1) x (C - 1) = (3 - 1) x (3 - 1) = 2 x 2 = 4

critical values: $\chi^2_{(4).05} = 9.49$; $\chi^2_{(4).01} = 13.28$

Since the calculated value $\chi^2 = 18.87$ is than the critical value $\chi^2_{(4).01} = 13.28$, reject H_0 at $p < .01$

Children's ability to recall words was influenced by different levels of environmental distractions.

12. using the formula for computing the theoretical values, $T = \frac{RxC}{G}$, and maintaining the same marginal frequencies as were observed, we obtain the following table of theoretical values:

	Support	Do Not Support	
Freshmen	31.25	68.75	100
Sophomores	31.25	68.75	100
Juniors	31.25	68.75	100
Seniors	31.25	68.75	100
	125	275	400

$$\chi^2 = \Sigma \frac{(O-T)^2}{T}$$

$$= \frac{(25-31.25)^2}{31.25} + \frac{(75-68.75)^2}{68.75} + \frac{(25-31.25)^2}{31.25} + \frac{(75-68.75)^2}{68.75}$$

$$+ \frac{(35-31.25)^2}{31.25} + \frac{(65-68.75)^2}{68.75} + \frac{(40-31.25)^2}{31.25} + \frac{(60-68.75)^2}{68.75}$$

$$= 1.25 + 0.57 + 1.25 + 0.57 + 0.45 + 0.20 + 2.45 + 1.11$$

$$= 7.85$$

df = (R - 1) x (C - 1) = (4 - 1) x (2 - 1) = 3 x 1 = 3

critical values: $\chi^2_{(3).05} = 7.81$; $\chi^2_{(3).01} = 11.34$

Since the calculated value $\chi^2 = 7.85$ is than the critical value $\chi^2_{(4).05} = 7.81$, reject H_0 at $p < .05$

Students from different class levels differentially support stricter eligibility requirements to remain enrolled at the university.

13.
a. multiple-sample χ^2: the data is nominal-level (categorical) and there are two variables, sampling (gender) and categorical (type of conflict strategy)
b. single-factor ANOVA; exam scores is ratio-level data and there is one independent variable (number of hours of TV viewing per day) with three levels
c. multiple-sample ANOVA; number of statements is ratio-level data and there are two independent variables, gender of participant and gender of partner
d. log-linear analysis: the data is nominal-level and there are three variables, gender, size of high school, size of college
e. t-test; the data is interval-level and there is one independent variable, purchasing location, with two levels (catalogue, store)